INFORMATION AND COMMUNICATIONS TECHNOLOGY AS POTENTIAL CATALYST FOR SUSTAINABLE URBAN DEVELOPMENT

T0133783

The European Institute for Comparative Urban Research, EURICUR, was founded in 1988 and has its seat with Erasmus University Rotterdam. EURICUR is the heart and pulse of an extensive network of European cities and universities. EURICUR's principal objective is to stimulate fundamental international comparative research into matters that are of interest to cities. To that end, EURICUR coordinates, initiates and carries out studies of subjects of strategic value for urban management today and in the future. Through its network EURICUR has privileged access to crucial information regarding urban development in Europe and North America and to key persons at all levels, working in different public and private organisations active in metropolitan areas. EURICUR closely cooperates with the Eurocities Association, representing about 80 large European cities.

As a scientific institution, one of EURICUR's core activities is to respond to the increasing need for information that broadens and deepens the insight into the complex process of urban development, among others by disseminating the results of its investigations by international book publications. These publications are especially valuable for city governments, supra-national, national and regional authorities, chambers of commerce, real estate developers and investors, academics and students, and others with an interest in urban affairs.

Euricur website: http://www.euricur.nl

This book is one of a series to be published by Ashgate under the auspices of EURICUR, European Institute for Comparative Urban Research, Erasmus University, Rotterdam. Titles in the series are:

Urban Tourism
Leo van den Berg, Jan van der Borg and Jan van der Meer

Metropolitan Organising Capacity
Leo van den Berg, Erik Braun and Jan van der Meer

National Urban Policies in the European Union
Leo van den Berg, Erik Braun and Jan van der Meer

The European High-Speed Train and Urban Development
Leo van den Berg and Peter Pol

Growth Clusters in European Metropolitan Cities
Leo van den Berg, Erik Braun and Willem van Winden

Information and Communications Technology as Potential Catalyst for Sustainable Urban Development

Experiences in Eindhoven, Helsinki, Manchester, Marseilles and The Hague

LEO VAN DEN BERG
WILLEM VAN WINDEN

European Institute for Comparative Urban Research
Erasmus University Rotterdam
The Netherlands
www.euricur.nl

LONDON AND NEW YORK

First published 2002 by Ashgate Publishing

Reissued 2018 by Routledge
2 Park Square, Milton Park, Abingdon, Oxon, OX14 4RN
711 Third Avenue, New York, NY 10017

Routledge is an imprint of the Taylor & Francis Group, an informa business

Publisher's Note
The publisher has gone to great lengths to ensure the quality of this reprint but points out that some imperfections in the original copies may be apparent.

Disclaimer
The publisher has made every effort to trace copyright holders and welcomes correspondence from those they have been unable to contact.

A Library of Congress record exists under LC control number: 2001096522

ISBN 13: 978-1-138-72355-9 (hbk)
ISBN 13: 978-1-138-72350-4 (pbk)
ISBN 13: 978-1-315-19295-6 (ebk)

Contents

List of Figures

List of Tables

Preface

In 1999, the European Institute for Comparative Urban Research (EURICUR) was invited by members of the Technology Policy Forum of Eurocities to carry out an international comparative study on technology and urban development.

During the investigation period, the Dutch Minister of the Interior and Kingdom Relations decided to organise a large international conference about roughly the same theme: ICT and the city. It was decided that the city of The Hague and the Ministry would team up in the organisation of the conference. The findings of the Euricur study would form a perfect input.

This book contains the findings of our investigation. It sheds new light on the impact of ICT on urban development and offers scope for local and regional governments to pursue effective policies. The main part of the information needed for the cluster analysis was generated by interviewing representatives of municipalities, firms, educational institutes and other organisations in the cities of Eindhoven, Helsinki, Manchester, Marseilles and The Hague. Without their help and information, the investigation could not have been accomplished at all. Therefore, we want to express our great appreciation for the welcome we received in each of the participating cities.

The research has been carried out under supervision and guidance of the city of The Hague, represented by Mr André van der Meer. We thank him warmly for his commitment, his enthusiasm, his valuable contributions and, last but not least, his 'organising capacity', which proved indispensable to carrying out a project with so many cities and people involved.

Next, we would like to thank the representatives of the participating cities. They were of great help in the organisation of the visits and the feedback of the results to the discussion partners, and contributed with valuable remarks. We thank the following people for their pleasant cooperation (in alphabetical order): Mr Jean-Claude Aroumougom (Marseilles), Mr Leo Draisma (The Hague), Mr Dave Carter (Manchester), Mr Eero Holstila (Helsinki), Mr Vesa Kanninen (Helsinki), and Mr Peter Nagel (Eindhoven).

We thank our colleague Alexander Otgaar for his contribution to the case of Helsinki. Naturally, only the two authors are responsible for the contents of the full book. Finally, we should like to thank Ms Attie Elderson-de Boer for screening the English language, and the Euricur secretary Ankimon Vernede for her indispensable help during the investigation.

Leo van den Berg
Willem van Winden

PART ONE
INTRODUCTION AND
CONTEXT

Chapter One

ICT and the City: An Introduction

1   Introduction

At the dawn of the twenty-first century, information and communications technologies (ICTs) are at the centre of interest for both businesses and governments. Increasingly, urban policy makers are also concerned with the new developments, opportunities and threats offered by digital revolution, and feel the need to respond strategically.

Usually, technology policy is not associated with the urban level but rather with the national or European level. Many types of technology policy do indeed take place at these levels: European and national technology policies are concerned with the stimulation of research and development (R&D) in large and small firms or in specific sectors such as commercialisation of research, technology adoption, standardisation, legal issues, property rights and so on. However, cities can pursue technology policies as well, as a means to reach economic and social objectives, as well as objectives intended to raise the quality of life and improve the internal and external accessibility of the urban region.

To shed more light on the impact of ICT on urban development and the consequences for urban management, the European Institute for Comparative Urban Research has been invited by the city of The Hague to carry out an investigation into the 'state of the art' concerning the use of information and communication technologies in five European cities. This book contains the results of an analysis of the impact of new technology and new technology policies in the cities of Eindhoven, Helsinki, Manchester, Marseilles and The Hague.

This chapter introduces the themes that will be discussed in more detail in the book. Section 2 elaborates on the aim of the research project, explains the methodology that was used, and elaborates on the case studies. Section 3 introduces the case studies of Eindhoven, Helsinki, Manchester, Marseilles and The Hague. Section 4 shows how the rest of the book is organised.

2   Aim and Methodology

This book has the ambitious aim to help to understand how ICT affects urban development and change. More specifically, we try to:

- understand the role of cities as attractive places for ICT companies;
- understand how the deployment of ICT can make a city more attractive in general;
- describe and analyse examples of good practice in urban ICT policies in several European cities;
- identify opportunities and pitfalls regarding ICT policies, based on experiences in the case-study cities;
- provide lessons for other cities and learn what urban managers and other decision-makers can do to make optimum use of the new technologies.

Much use has been made of existing literature, particularly of earlier Euricur studies on growth clusters in European metropolitan regions (Berg, Braun and Van Winden, 1999). Fieldwork, however, yielded the key input for this book. Most of this book's content is the result of an explorative study into the way European cities deal with ICT developments. We investigated five European cities: Eindhoven, Helsinki, Manchester, Marseilles and The Hague. For each city involved we reviewed the available reports and studies on the subject thoroughly. On that basis, we were able to identify and select key actors. After that, we visited the case cities and interviewed the selected persons.

3   Introduction to the Cases

The cities included in our study are interesting for our purposes for several reasons. First (although they are not the only ones in Europe), each of the cities has an explicit ICT strategy and seeks to translate that strategy into concrete actions. In all of the cities, ICT is a policy spearhead. Second, each city has a different emphasis in their policies and deploys ICT for different purposes and in different ways: this helped us to understand and describe the wide range of possible ICT policies. Third, the case-study cities are located in four different European countries. This European perspective enriches the study, as it reveals similarities and differences in ICT development and policy opportunities in several national contexts. Table 1.1 shows the focus of our separate cases. Below, their content is briefly described.

Table 1.1 ICT and the city: 5 case studies

City	Topic
Eindhoven	ICT and future living: e-city, an ICT experimental neighbourhood
Helsinki	ICT and inner city development: Lasipalatsi
Manchester	ICT cluster development and policies
	ICT adoption: reducing the digital divide
Marseilles	ICT and public service delivery
The Hague	ICT cluster development and policies
	ICT and public service delivery

Eindhoven

In the Eindhoven region, we studied the development of the 'e-city'. In this public/private project, 80,000 citizens have broadband connections and access to the Internet. Firms and local government develop electronic services for the population in this experimental zone. The experiment enables companies to offer services on a sufficient scale and thus speeds up the transition to the information society. Many of the experiences may be applied on a large scale later on. The case shows how a region can build a lead by creating conditions for experimentation.

Helsinki

This case study centres around an innovative project in Helsinki named Lasipalatsi ('glass palace'), a media centre in the inner city of Helsinki. The public building serves to disseminate innovations in the ICT and new media, bring technology closer to the people, and improve contact between technology developers and users. The combination of these functions in a leisure-like setting is new for Europe. For the city of Helsinki, the specific public/private organisational form of the project is innovative; additionally, for Helsinki the project was the first large project in which European money was involved.

Manchester

In this case study, we first focus on how ICT offers new opportunities for both economic and social regeneration, and how urban management seizes them. Furthermore, we describe and evaluate regional policies to further the ICT cluster. Second, we address the question how ICT can contribute to the inclusion of less favoured groups into the information society, and show some examples of projects in Manchester in this field.

Marseilles

This case study shows how the municipality uses new technologies to improve service delivery to its population. Some examples of ICT projects are presented, showing not only how ICT can contribute to the solution of problems and the seizing of new opportunities, but also what technical and organisational difficulties may arise in these renewal processes.

The Hague

The City of The Hague has the ambition to become a strong centre of ICT and telecom activity. In this case study, we analyse the state of the art of the ICT-cluster in the region: what ICT-related actors can be discerned in the Haaglanden region, how they interact and what their location considerations are. Also, we assess and judge the policies and ambitions of the different public players in the region.

4   Organisation of the Book

Figure 1.1 shows how this study is organised. Each chapter can be read separately. Chapter Two sketches the context of this study: what is happening in urban Europe, what are the threats and challenges that cities face, how can cities react strategically? Finally, in this chapter we introduce a conceptual framework that helps to understand how new information and communication technologies affect urban development, and how urban management may respond strategically.

Chapters Three and Four present and analyse the core findings of the study. Chapter Three focuses on the potential role for ICT firms as new engines of economic growth in urban regions in Europe and provides strategic lessons for cities. The subject of Chapter Four can be summarised as 'ICT and urban attractiveness': how can ICT be deployed as a tool to improve the urban product and its accessibility? What are the technical possibilities, organisational pitfalls and bottlenecks and what is the state-of-the art in European cities? In Chapter Five, we bring the themes together, draw conclusions and suggest topics for further research. The separate case studies can be found in Chapters Six to Ten.

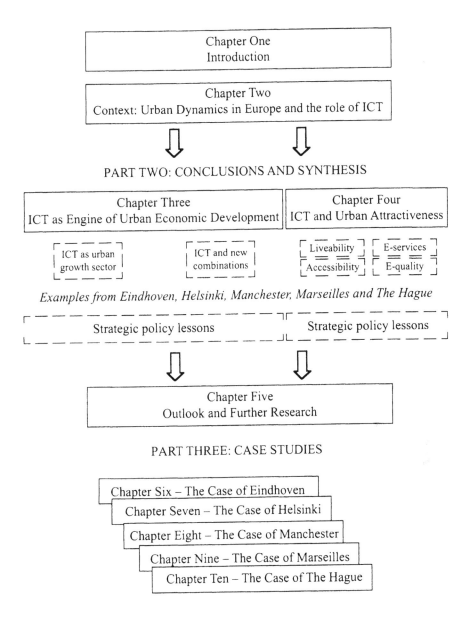

PART ONE: INTRODUCTION AND CONTEXT

Chapter One
Introduction

Chapter Two
Context: Urban Dynamics in Europe and the role of ICT

PART TWO: CONCLUSIONS AND SYNTHESIS

Chapter Three
ICT as Engine of Urban Economic Development

Chapter Four
ICT and Urban Attractiveness

ICT as urban growth sector

ICT and new combinations

Liveability   E-services
Accessibility   E-quality

Examples from Eindhoven, Helsinki, Manchester, Marseilles and The Hague

Strategic policy lessons

Strategic policy lessons

Chapter Five
Outlook and Further Research

PART THREE: CASE STUDIES

Chapter Six – The Case of Eindhoven
Chapter Seven – The Case of Helsinki
Chapter Eight – The Case of Manchester
Chapter Nine – The Case of Marseilles
Chapter Ten – The Case of The Hague

Figure 1.1   Organisation of the book

Chapter Two

Context: Urban Dynamics in Europe and the Role of ICT

1 Introduction

Although the impact of ICT on urban development is pervasive and all-embracing, it is also part of other, more general developments that occur in Europe's urban regions. In this chapter, we sketch some major developments affecting urban Europe, as a context for our analysis. After this contextualisation, we develop a conceptual framework that helps to understand the role of ICT in urban development. This is not an all-encompassing theory, but rather a fruitful means to analyse the role of ICT in the improvement of the well-being of citizens and companies in a city, and a useful tool to derive ICT policy options for urban management.

This chapter starts, in section 2, with a more general analysis of the dynamic developments that affect urban regions throughout Europe: globalisation, localisation, informationalisation, the rise of polycentric regions and the growing urban competition. This analysis leads to the conclusion that urban regions need to improve their attractiveness in order to be socially and economically successful in the long run.

In section 3 the notion and importance of urban attractiveness is further elaborated and conceptualised. The focus then shifts to the question of how the information revolution affects the attractiveness of a city for citizens, companies and visitors. On this basis, we formulate the scope for urban ICT policy in section 4. Section 5 concludes.

2 Mega Trends

Urban dynamics is determined by certain trends that manifested themselves in Europe halfway through the 1980s (Braun and Van der Meer, 2000). Globalisation has fundamentally influenced the international urban development process and thereby also the competitive positioning of cities (Hall, 1995). More and more commercial activities are looking at the world as a whole when making location choices, while national borders seem to play an ever-decreasing role (Ohmae, 1995). Companies can, in this way, expand their markets, but will at the same time experience more competition in their home market. How far this is favourable for the region depends on the regional competitive capacity. The strength of this competitive capacity, in turn, depends on, amongst other things, the international orientation, the size of the regional commercial activity, the presence of decision centres in the region and the presence of strong economic clusters with growth prospective. Globalisation appears to manifest itself increasingly on a local level: the success of cities and regions seems to be based on local factors and institutional structures and the use that strategists are able to make of them (Swijngedouw, 1999).

> Paradoxically the enduring competitive advantages in a global economy lie increasingly in local things – knowledge, relationships, and motivation that distant rivals cannot match ... What happens inside companies is important but clusters reveal that the immediate business environment outside companies plays a vital role as well (Porter, 1998).

Globalisation has become possible through information and communication technology. The transition to an information society has considerably strengthened the position of cities as nerve centres of the 'New Economy'. Cities provide the daily context for the increasingly global and footloose interactions within the economic, social and cultural spheres. The centres of large cities are historically the locations where information was made known and exchanged. Because the need for face-to-face communication barely seems to be suffering, it is expected that, in Europe, the large cities, and within them the city centres, will retain that role. Also, city regions offer the agglomeration advantages (varied and specialised activities and facilities, qualified and varied workforce) and the quality of life (excellent facilities, ambience, vibrancy, art and culture) while the large cities are the hubs of the international transport infrastructure.

> The new locational logic is governed by access to information. This is obtained either by face-to-face communication or by electronic transfer. As telecommunication costs have dropped considerably informational activities should have been increasingly free to locate away from old face-to-face exchange locations. But evidently they have not ... (Hall, 1995).

The disadvantages of core cities are the usual problematic internal accessibility, the shortcomings of the quality of living space and the increasing social polarisation and insecurity. The suburban areas compensate for these disadvantages, which is why there is so much interest in establishing living and working areas in or close to large cities.

Globalisation and informationalisation do not only contribute to important internal changes in city regions (such as social, economic and spatial polarisation); the relationships between cities, and the position of cities within international urban systems are also changing (Hall, 1995). Kunzmann speaks of an increasing spatial specialisation with consequences such as spatial differentiation, growing spatial polarisation and unavoidably further growth of the urban region in Europe, expanding more and more into a wider hinterland (Kunzmann, 1996).

These mega trends bear the stamp of urban competition. Increasingly, cities and towns behave in a logic of competition in a highly dynamic and complex environment (Bramezza, 1996). In such a competitive environment the old 'certainties' no longer exist: although the geographical situation is still relevant it is no longer as dominant as it used to be. Qualitative location factors have come to be very important. That gives cities an incentive to invest in their own attractiveness. But it is not enough for a city to be more attractive than its competitors as a location for enterprise. The inhabitants, the businesses, investors and visitors determine whether or not a city is attractive. These (potential) customers of cities put high demands on the quality of the business, living and visiting environment (Braun and Van der Meer, 2000).

3   ICT, Urban Growth and Urban Attractiveness: A Conceptual Framework

This brings us close to the central questions of this study: in what ways does ICT affect the attractiveness of the city as a place to live or work, or for companies, as a good place to locate, or, for tourists, as a good place to visit? How can cities benefit from the growth of the ICT sector? And, what can urban management do to improve the relative attractiveness by using ICTs?

Before the last question can be answered, we need to develop a conceptual framework to assess the way ICT changes the attractiveness of an urban region. A fruitful conceptual way is to define attractiveness in terms of accessibility of *welfare elements* for the population, for firms and for visitors (Van den Berg, 1987). The attractiveness of a city for the population depends on the accessibility of shops, leisure facilities, job opportunities, health care, quality housing provision, safety and so on.

The attractiveness of a city for firms depends on the accessibility of appropriately skilled labour, inputs – not only physical inputs, such as components, but also non-material inputs such as business services or partner firms – and good business locations. For tourists or visitors, what counts is the accessibility of tourist attractions, events, cultural facilities, hotels, etc.

It is crucial to distinguish between the *presence* of welfare elements on the one hand and their *accessibility* – costs and efforts in terms of time, money, and all possible efforts to reach them – on the other. For instance, if a shopping centre is hardly accessible for lack of parking space or lack of transport connections, the centre is much less valuable to the citizens than if it were very easily accessible. The welfare of the population can be improved by improving the centre's accessibility. Similar examples can be given for companies and tourists.

For citizens, a city is more attractive the more welfare elements are accessible. The same holds for tourists and tourist attractions/leisure facilities respectively. If we think of urban attractiveness in this way, we can clearly assess the role of ICT. *Information and communication technologies change both the nature of welfare elements, and their accessibility* (see Figure 2.1).

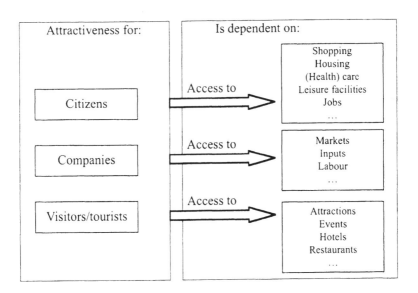

Figure 2.1   Urban attractiveness and welfare elements

ICT Changes the Accessibility of Welfare Elements and Location Factors

ICT has the capacity to break down barriers of space and time. As such, it makes welfare elements (for citizens) and location factors (for firms) better accessible (see Figure 2.2). For some functions, ICT has changed the world into a 24-hour marketplace. For instance, online shopping enables customers to shop in the middle of the night at a shop 'located' in the United States. It brings within reach all kind of products and services they would never have been able to buy before. This improves their welfare perception. For citizens who are looking for a job, an online job intermediary greatly improves their chances of finding the job that they want. In other words, the accessibility of employment is improved.[1] The accessibility of health services can be improved enormously by introducing teleconsulting services: disabled people no longer have to travel long distances to hospital for a simple diagnosis. In leisure, the accessibility of events, concerts, attractions etc. can be greatly improved if ICT is used to provide citizens with customised information and online reservation possibilities. It brings within reach a broader range of leisure services, greatly reduces search costs and effort and contributes to the welfare perception of the citizen or the tourist.

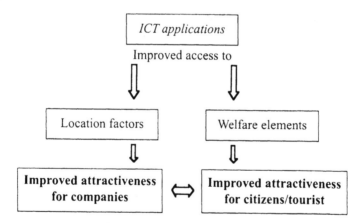

Figure 2.2   ICT and the accessibility of welfare elements/location factors

Likewise, the Internet makes it easier for companies to enlarge their market reach, or, to put it differently, the accessibility of markets is greatly improved. Goods can be sold to customers all over the world by using the Internet as distribution channel. Not only the geographic scale of markets is increasing with the use of ICT: entire production processes can be organised more efficiently, making use of local conditions. For instance, a car producer can outsource the production of parts to a remote low-wage country. He can still control the process, by sending exact specifications of parts over the net, and can track and trace the production and transport in real time. As such, ICT greatly speeds up the process of internationalisation and leads to new location patterns of businesses, as well as to different development dynamics of regions that have different endowments.

The examples above all concern the creation of virtual accessibility. However, ICT can also be deployed to improve physical accessibility, for instance when real time traffic information is provided at a website, or when traffic management systems are used to reroute road traffic flows in cases of congestion, and so on.

ICT Changes the Nature and Quality of Welfare Elements

It would be one-sided to regard ICT only as distance-shrinking technology: it is much more than that. ICT may not only improve the accessibility of existing welfare elements: it can change the nature – or improve the quality – of welfare elements as such. One aspect is the ability of ICT to gear existing products and services to individual needs. For instance, online journals or newspapers can offer just the information that the reader is interested in rather than the whole content. ICT also enables more radical new forms of existing products or services. In health services, for instance, ICT makes it possible to create virtual teleconsulting: you no longer need actually to go to the doctor for any complaint: part of the doctor's activities (the diagnosis) can be done at home. More sophisticated examples could be remote monitoring of patients at home using teleconference technologies. The patient no longer needs a hospital bed but can stay at home; only in cases of emergency is a doctor called in. In other words, medical activities are split up into parts, some of which are organised in a new way. The end result might be better care, improved efficiency and considerable cost reductions.

ICT also yields new products and services that exist only in cyberspace. Instances are virtual products without a physical reference, such as interactive games. Another example is the formation of virtual communities. For many people, this offers new possibilities of communication which enhances their welfare perception.

4   The Role of Urban Management

Ways in which ICT can improve the accessibility of welfare elements and thus contribute to the welfare of citizens and companies and hence to the attractiveness of a city have been outlined above. What is the role of urban management in this respect? A first important remark is that in the majority of instances, the market functions well and 'equips' citizens with new opportunities without government intervention. For instance, retail business invests heavily in electronic shopping services; banks provide online banking services to everyone who has a PC. Despite the strong dynamics of the market, (local) governments might play a role in several respects.

* *Improve access.* It is a fact that in most European cities, some people – many of them from the socially weaker groups – have no access to the Internet: they have no computer at home, because they lack either the money or the skills to use it. This implies that they have fewer welfare elements at their disposal. If urban management manages to equip them with skills and equipment – in cooperation with the private sector – the welfare perception of the population would be greatly enhanced. It may also result in a critical mass of ICT users, which in turn may make it very interesting for companies to offer new services. This raises welfare levels even further. It is critical not only to provide access but to link it with e-literacy programmes.
* *Improve the supply of public services.* Public organisations and administrations can use ICT to raise their service levels. For instance, citizens may renew their passports online, or get specific government information when needed. E-government not only improves customer satisfaction and saves money for the government: if a city manages to become very efficient and accessible, it improves its attractiveness as a business location as well.
* *Improve physical accessibility.* Urban management has a powerful role in intra-urban transportation matters. It can improve the internal accessibility greatly by using ICT, for instance by (co)investing in traffic information systems, online public transport booking services, resulting in improved accessibility of welfare elements for citizens, firms and visitors.
* *Speed up the development of ICT applications in the (semi-) public domain.* In some sectors – the health sector is a good example – adoption of ICTs to improve service delivery is a slow and difficult process. Reasons can be the lack of incentives, lack of resources, or the complexity of transactions.

Local government can play several roles in the acceleration of ICT service provision, for instance as co-funder of pilot projects, or network broker (bringing actors together).

For all the roles of urban management, it holds that the private sector should be involved, as a source of funds and expertise.

The possibilities of ICT to improve urban attractiveness are almost endless, but resources – in terms of money and effort – are scarce. This implies that priorities have to be set for urban ICT policy. From a general urban welfare point of view, it is best to invest in ICT solutions with the highest marginal return, i.e. those that solve the most urgent problems and/or are relatively cheap. Given the diversity of urban problems, a different ICT policy mix should be applied for each city. Once urban management has decided in what field or function ICT policy should be deployed, new questions arise concerning the design and implementation of policies. In the chapters that follow, we will pay attention to this matter.

5   Conclusions

Fundamental economic and technological developments set the stage for new urban dynamics. Cities are confronted with new opportunities and threats, and should act strategically. The emergence of information and communication technologies is perhaps the most powerful force in that respect. Urban attractiveness are key words in the urban competition for firms, companies and inhabitants.

In this chapter, we have introduced a conceptual framework that helps to think about the relation between ICT and urban development and refine the notion of urban attractiveness. In short, we have pointed out that ICT can improve the attractiveness of an urban region – for citizens, firms or visitors – because it makes existing welfare elements better accessible, or creates entirely new forms of welfare elements. This framework is not an all-encompassing theory, but rather a fruitful means to analyse the role of ICT in the improvement of the well-being of citizens and companies in a city, and a useful tool to assess ICT policy of urban management. It certainly needs more elaboration.

In the chapters that follow, we will apply the frame to two basic themes, and elaborate on the organisational issues of technology policy. The first theme is attractiveness of urban regions as a location for ICT companies. We will discuss how urban management may improve their city's attractiveness for these types of firms. The ICT sector is a booming sector and an important source of new wealth for cities. In Chapter Three we elaborate this theme, showing experiences in several European cities.

The second theme to be addressed is more general: we will assess how urban management can deploy ICT-policy to improve the attractiveness of an urban region. We focus on ICT and urban transportation, ICT and public services delivery and the provision of access to the Internet for all citizens. Also in this chapter, cases are illustrated with examples from several European cities.

Note

1   In some cases, the job market even becomes global: some US software developers employ Indians (who still live in India!) to develop software modules. All the contacts are electronic.

PART TWO
ICT AND THE CITY: CONCLUSIONS AND SYNTHESIS

Chapter Three

The ICT Sector as Growth Engine for Urban Regions

1   Introduction

ICT is the principal growth sector of the early twenty-first century. Its development is driven by accelerating technological evolution, resulting in a continuous stream of new products and applications on the market, eagerly used by companies and consumers. An important growth impulse for the ICT-sector is the liberalisation of telecom markets in Europe, which entails new players, lower prices, more competition and a much accelerated innovation process.

For cities, the growth of this sector opens perspectives for new economic development, wealth creation and employment growth. What are their chances to benefit, which factors are critical, and what should cities do to profit optimally from the sector's growth? These are the central questions of this chapter.

The chapter is organised as follows. In section 2, we start with a discussion of the definition and composition of the ICT sector, because there is much unclarity about this matter. Then, in section 3, we list the principal trends that affect the sector. A distinction is made between market, technological and institutional trends. Section 4 forms the core of the chapter. Here, the questions of why and how urban regions would benefit from the information revolution are addressed. Section 5 gives an overview of possible urban policies to boost the ICT sector, enriched with many examples from our study in five European cities. Section 6 concludes and derives strategic lessons for cities.

2   The ICT Sector: Definitions

Before discussing the impact of ICT on the urban economy, we should clarify what is meant by the ICT sector. Unfortunately, there is no generally accepted definition. Rather, a number of distinctions and definitions coexist. We will discuss some of them.

The Dutch National Institute for Statistics (CBS) draws a distinction between ICT manufacturing and ICT services (CBS, 1999). ICT manufacturing consists of all (sub)sectors where goods are produced that contribute to the electronic infrastructure for both communication and the provision of information. Examples are producers of: office machinery and computers; wires and cables; electronic components; telecommunication equipment. ICT services consist of all (sub)sectors that supply services in the fields of telecommunication and information technology that make the utilisation of the (electronic) infrastructure possible (CBS, 1999). Both OECD (2000) and CBS (1999) place the ICT manufacturers and ICT services in a broader reality that is called the information economy or the information society. Besides the producers of the electronic infrastructure and the service providers who make use of the electronic infrastructure, a third group of companies is distinguished: the suppliers of information products or 'content'. Examples are broadcasters, publishers and advertisement agencies. There is not yet consensus about the exact definition of the 'content'-sector. An important point of discussion is where to draw the lines within the wide range of entertainment, arts and culture related activities.

Other institutes use more refined distinctions, based on business processes. A recent example is a Booz, Allen & Hamilton (2000) study, in which the following subsectors are discerned:

- IT service providers and software developers. The software sector can be subdivided into packaged and embedded software. The former contains applications like word processing and games and application tools like spreadsheets and system infrastructure software. Embedded software is the software that is already incorporated in all kinds of devices like mobile phones and personal organisers, at the point of manufacturing;
- providers of communication and network services, both fixed and mobile, which get the information to the end-user. Examples are telephone service providers;
- producers of the public network equipment (transmission and switching equipment, routers and cable producers): they produce the public electronic infrastructure. Examples are CISCO and Ericsson;
- components and end-user equipment. Here we find the commodities that link the end-users to the information society. The components subsector can be further divided into active components (such as integrated circuits, tubes and LCDs) and passive components (such as capacitors, connectors and resistors). Mobile and fixed private network equipment includes devices

such as telephone sets, mobile phones, television and radio transmission apparatus, switchboards and fax machines. IT hardware consists of office machinery, computers and other types of information-processing equipment.

Some other institutes define a specific Internet and e-commerce sector. Banks and technology funds, for instance, identify infrastructure companies, Internet service providers, business to business e-commerce, business to consumer e-commerce, networking, security, investors, portals and advertisers (NRC, 2000).

In the following sections, we will take a broad perspective on the sector. We will include traditional sectors such as telecom companies and equipment producers, as well as newer firms such as software developers, ICT consultants and Internet companies, and also electronic content providers. Additionally, we will pay generous attention to the 'urban demand side' for ICT products and services.

3   Trends

The heterogeneity of the ICT sector makes it difficult if not impossible to list an exhaustive set of trends in 'the ICT sector': each subsector is, to a large extent, subject to specific trends and developments. Nevertheless, some 'fundamental' trends can be listed that affect virtually all segments of the sector. A distinction can be made between market trends, technological trends and institutional trends.

Market Trends

- Growing market penetration of Internet. Internet use is expected to rise quickly, especially in Europe. That gives an enormous impulse for e-commerce activities; it is an incentive for even faster development of Internet applications, and other software products (interactive games, to mention only one), because the market grows by the minute.
- Exploding demand for telecommunications products and services. Drivers are the increased Internet penetration, increased volumes of data transport and the introduction of new services such as mobile Internet. New higher capacity infrastructure is developed to meet this demand.
- Falling prices in formerly regulated markets. Telecom services in particular have become much cheaper as companies have to fight for market share; they will continue to fall.

- Ongoing mergers, acquisitions and strategic alliances, some with the aim to exploit synergies, others to create scale. The Time Warner–AOL merger in 1999 is an example of an effort to combine content production (Time Warner) and distribution (AOL). The merger trend leads to large companies, sometimes with substantial market power, but also contributes to the internationalisation of the ICT sector.
- Increased availability of venture capital. Finding risk capital used to be a difficult exercise, with banks reluctant to lend. Following the American trend, venture capital is now abundant in Europe. This enables high tech firms to engage in new projects, and spurs the number of start-ups. Figures released by the European Venture Capital Association reflect the explosion of activity in Europe's start-up scene: investments made in early stage high-tech companies quadrupled in 1999 compared with 1998 to EURO 16 bn (*Financial Times*, 7 June 2000).

Technological Trends

- *A fast pace of technological development.* Technology developments are very rapid in all segments of the market. Therefore, ICT companies need to invest in R&D, or make sure to be *au courant* with the latest trends. Time spans of investment projects are very short: uncertainty rules.
- *Electronic infrastructure.* Many innovations increase the speed and capacity of existing data transportation (such as telephone lines) and new infrastructures, such as GPRS, UMTS and digital broadcasting technology, are being developed. High-capacity infrastructure enables the development of all kinds of new services.
- *The mobile revolution.* Mobile telephones have become ubiquitous in a few years. In the next few years, mobile Internet will probably take off. Furthermore, wireless applications can be used in many situations, for instance in cars (see box). Blue tooth technology enables computers to connect with each other, offering a range of new possibilities.
- *Embedded technology.* Increasingly, information technology is incorporated in conventional products. One example is the use of telematics in cars: see box below. Mitchell (1999) provides a long list of other examples of 'smart products', e.g.: clothes with built-in sensors that warn you when you need a shower; windows that can be opened/closed with your PC; intelligent fridges that automatically order new products when you take something out of it, etc.

Telematics in the car industry

Telematics has been a hot topic in the automotive industry, with most manufacturers seeing huge potential for offering value-added services – from email access to navigational assistance – as wireless technology improves. Some estimates have suggested that half the new vehicles sold in North America, Japan and western Europe could have 'embedded communications' modules by 2005. Ford Motors, the second largest car and truck maker, plans to install telematics capabilities – for providing in-vehicle information, Internet and entertainment services – on virtually all its new vehicles by the end of 2004 and on about 3m by 2003. Ford is teaming up with Qualcomm, the US wireless chip company, to form a jointly-owned company, Wingcast, to spearhead its telematics initiatives.

Source: *Financial Times*, 31 July 2000.

Institutional Trends

- *Market liberalisation.* Less than a decade ago, most European countries had only one supplier of telecom services: the state monopolist. This landscape has changed drastically. Under the strong influence of European policies most countries have freed their market to some extent; since then, many entrants have entered the various national markets. However, there is no 'level playing field': each country has its own pace of liberalisation and many regulations are still of a national nature; some former state monopolists are in a more favoured position than others in their home market, which enhances their chances of expansion abroad. This hampers the sector's development, as scale economies cannot be exploited on a European scale.
- *Deregulation and reregulation.* In Europe during the last decade a number of restrictions have been removed, paving the way for a freer telecom market. New regulations have been developed to safeguard competition and prevent the formation of monopolies. However, there is still no European competition agency dedicated to the telecom sector.
- *Intellectual property rights.* The ease with which content can be downloaded and copied raises urgent questions on how to safeguard intellectual property rights. Recent legal action in the US against Napster (a site that helped to access music files) shows that the debate has only just begun.

4   The Attractiveness of Urban Regions for ICT Firms

The ICT sector is highly dynamic: institutions, markets and technologies are evolving rapidly. The central theme of this section is the extent to which urban regions benefit from the explosive growth of the ICT sector. In the literature there is support for the thesis that urban regions profit more than proportionally from the rise of ICT. Many authors (Abler, 1977; Alles et al., 1994; Graham and Marvin, 1996; Schmandt et al., 1990, Shields et al., 1993) find that both the development and application of ICTs (infrastructure, hardware and software) do indeed occur in or near urbanised economic core regions. The best known examples are Silicon Valley in California, the M29 corridor near London and Route-128 near Boston in the USA.

We suggest that ICT can be an engine behind urban economic growth in two ways. First, cities can benefit from the location of the new and fast-growing ICT companies, which generate employment, added value and purchasing power. Second, cities may be the locations for emerging fruitful marriages between the ICT sector and 'old' sectors, or, to put it differently, the place where 'new combinations' emerge.

Why would urban regions, rather than other areas, benefit from the development of the ICT sector? In other words: what does the attractiveness of urban regions for ICT activity consist of? In line with our conceptual framework, we define attractiveness in terms of access to relevant provisions. A number of factors can be discerned. First, cities are nodes on (international) infrastructure networks, a very important amenity in a networked sector like the ICT sector. Second, cities provide access to concentrations of human resources, the most important resource for ICT firms. Third, cities facilitate access to markets and inputs: they form large concentrations of economic activity, which makes them attractive for ICT companies; and fourth, cities offer a quality of life that appeals to many ICT employees and firms. These arguments are elaborated below, enriched with examples from our case-study cities.

General Access: Cities as Infrastructure Nodes

The good accessibility of most urban regions contributes to their attractiveness as locations for ICT companies. More than other sectors, the ICT sector is a very flexible and 'networked' sector: companies often cooperate with other (ICT) companies in projects, in ever-changing alliances. Accessibility is therefore critical: ICT companies often prefer locations near highways, railway

They provide:

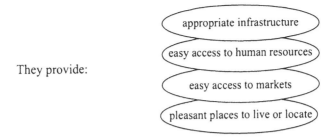

Figure 3.1   Why are urban regions attractive to ICT firms?

stations and, if they operate internationally (as many do), international airports. In Manchester the southern fringe of the city, close to the airport, is a very popular spot for ICT companies and hosts an expanding cluster. E-commerce firms in The Hague (the headquarters of Amazon.com and Mapquest) greatly appreciate the proximity of Schiphol international airport.

The quality of the *electronic* infrastructure matters as well, in some instances. Because cities are such concentrations of economic activity, they are best provided with electronic infrastructure (see box), which makes them in turn attractive to ICT firms.

MCI WorldCom: principal cities the first to be connected by glass fibre
The worldwide operating American telecom firm WorldCom (annual turnover: US$ 30 bn) is in the process of laying out a world-spanning glass-fibre network, which is to serve voice and data traffic. The company addresses business users who have to process masses of verbal, data and Internet flows, such as banks, call centres and commercial offices. In the USA the network has already branched out, inter-connecting all the major towns where the customers are. Europe was recently incorporated into that network by a transatlantic cable (laid out in a joint venture with Cable & Wireless; capacity: 30 gigabits a second). Amsterdam, London, Paris, Brussels, Düsseldorf and Frankfurt were the first European cities to be connected by a high-capacity fibre optic network, mainly because the principal customers are found in those towns. Local high-capacity networks have by now been constructed in more European towns, among them Stockholm, Zurich, Hamburg and Rotterdam.
Source: *Rotterdams Dagblad*, 1998; MCI-WorldCom homepage.

Some firms have special needs with respect to the quality of the information infrastructure; this particularly holds for companies that transmit enormous flows of data, such as Internet providers, electronic publishers or other media companies. Manchester can serve as example. Since 1998, TeleCity, the UK's only Internet exchange outside of London, has been located at the Manchester Science Park. This puts the area at the heart of UK Internet infrastructure development by offering the second major UK hub to the transcontinental Internet backbone (the other network access point in the UK is located in London). Businesses wanting to lease high-speed Internet connectivity can now take advantage of cheaper rates as access is direct rather than via London. This complements the Manchester Network Exchange Point (MaNAP) set up by the University of Manchester in 1997 and together these services have been a key factor in the rapid growth of the Internet and e-business sector in Manchester. In Eindhoven, the university campus is equipped with broadband infrastructure, and all students have a notebook that they can plug in anywhere. Several companies see this as an opportunity to experiment with new web-based applications.

Access to Human Resources

Big cities are concentrations of 'human capital', the most important resource for ICT companies. In many surveys, 'availability of skilled staff' is mentioned as one of the primary location factors (see for instance, MIDAS, 2000).

This point is illustrated in our case study of The Hague: two new telecom service providers in the Dutch market decided to locate in The Hague, one of the reasons being that this city offered a large pool of specialised tele-communications specialists, thanks to the presence of an incumbent firm, the former state monopolist KPN.

Another asset of most cities is formed by its knowledge institutes, notably universities. They not only supply the city with trainees and new graduates, but in some cases also execute valuable basic research in information science that may be of importance to the local ICT industry. Perhaps even more importantly, universities are important breeding grounds for new ICT firms, started by entrepreneurial students or researchers who commercialise their ideas. In the ICT sector, business start-ups play an important role; many of the now-large ICT companies such as Amazon.com were started from scratch by entrepreneurial youngsters in the 1990s. In the cities of Manchester and Eindhoven, this potential is clearly recognised. University based public/private companies offer space, legal and financial support to start-ups, and provide access to university facilities such as laboratories or powerful computers. In Eindhoven, the company takes a share in the start-up; in Manchester, start-ups have to pay royalties as a certain percentage of its turnover, after three years of profitable operation. In Manchester in the last five years, 50 university start-ups have been supported, offering 200 jobs (*Campus Ventures News*, 1998). Some of them are expanding rapidly.

Access to Demand for ICT Products and Services

The ICT sector is interwoven with other economic sectors: large parts of the ICT sector can be considered part of the business services sector. Cities are large concentrations of economic activity, which makes them attractive locations for ICT firms that prefer to locate near their clients. The 'heaviest' users of ICT are corporate headquarters, financial institutions and the media industry (Saxenian, 1994; Hall, 1998). The rapid development of online banking and new media (online publishing, web-TV and radio) and entertainment services in particular induce the ICT sector and the banking and media sectors to mix up. Financial centres are important ICT demanders: the financial sector has had the highest ICT expenses for several years now. This may explain the success of the cities of London, Amsterdam and New York in attracting ICT companies. The role of cities as concentrations of economic activity gives them a strong position to benefit from new combinations that arise from the application of ICT in existing sectors. Hall (1998) points at the fruitful marriages between ICT and typically urban sectors such as cultural, media and entertainment industries.

The integration of ICT with other sectors means that every city has chances to develop its ICT sector. It also suggests that each city develops a characteristic ICT sector with its own 'local colour', that fits the structure of the urban economy. The ICT sector in Munich for instance is very strong in media applications, triggered by the high concentration of media companies in that city (Berg, Braun and Van Winden, 1999). The ICT sector in the famous Silicon Alley in New York is dominated by content creation, and addresses the concentration of publishers in the city. In Manchester, the ICT cluster is not that specialised, although the local financial institutes and media companies (ITV and BBC) are the principal demanders of ICT services. In Eindhoven, the ICT sector is more linked to the industrial sectors (strongly represented in the region). ICT companies develop applications for order processing, logistics applications or e-business solutions, and network meetings are organised between old economy (industrial) and new economy (ICT) companies, to discover new combinations and business opportunities.

In some cities, the ICT sector is less interwoven with the other sectors, but plays a more independent role. The most outstanding example in Europe is probably Helsinki. The ICT cluster there is very large relative to other sectors and very internationally oriented. Its development is driven by the global player Nokia, but telecom service providers (such as Sonera) and small firms are also increasingly international. Many firms are active in R&D.

Cities as Pleasant Places

Labour shortage and rapid business expansion are the causes of increasing global competition for ICT talent. From this perspective, the quality of life in the urban region – not only the availability of comfortable housing, metropolitan ambiance, cultural conditions, green spaces, but also climate, etc. – is an important condition for attracting (or retaining) staff to the city and thus for developing a cluster. Mitchell's (1999) point that, in the end, the 'pleasant places' on earth will be the main ICT centres might be somewhat extreme, but there is much truth in it. In the Netherlands, the explosive growth of Amsterdam's ICT cluster can be partly explained by its young, dynamic and cultural image. Its vivid urban life – of course among many other factors – makes it an attractive place for (young) ICT entrepreneurs.

Helsinki's cluster has been booming for several years now, but the limits to expansion are in sight. Up till now, the city's 'knowledge base' has expanded, thanks to the immigration of young professionals from other villages and cities in the country. However, it is not easy to attract foreigners to the remote city, with its cold and dark winters. The strategy of many rapidly expanding ICT companies is now to attract workers from the Baltic. Manchester has also problems in attracting – and retaining – its talent. Although the city is known for its nightlife and has a large (and expanding) cultural scene, it faces heavy competition for talent from London. The capital acts as a magnet upon talented people in any sector, as well as for ICT professionals. Many students who are educated at one of Manchester's universities leave the town for a job in London. In The Hague, which is part of the Randstad, the situation is a little different. The city's ICT companies indicate no special problems in attracting ICT staff. This is not only due to quality of the city's environment, but also to the large labour market that is within reach: The Hague forms part of the Randstad area (with Amsterdam, Rotterdam and Utrecht), with five million inhabitants. A high percentage of the staff in The Hague-based ICT companies do not live either in the Hague or its region.

Most cities' policy makers recognise the importance of quality of life as general condition for knowledge-intensive activities such as the ICT sector. Helsinki invests heavily in new cultural facilities (the new museum of contemporary art is an example) and is upgrading the inner city area. The Hague is also upgrading and restyling its inner city, offering high quality housing and office space, and manages to attract highly educated people to the city. Manchester has a similar strategy: it seeks to improve the quality of its urban environment by restructuring the inner city and building expensive housing facilities, also in the city centre. The City of Eindhoven, although 'leading in technology' in The Netherlands, has problems with its image as a dull town. Some Eindhoven-based ICT companies try to benefit from Amsterdam's more dynamic image and present themselves as located 'very near Amsterdam', to impress foreign partners or potential staff (indeed, it takes only one and a half hours by train from Eindhoven to Amsterdam). The city council of Eindhoven has fully recognised the city's image problem for several years and invests much in cultural facilities, shopping malls, etc., but an image does not change overnight.

5   ICT Cluster Policies in European Cities

Almost every city wants to benefit from the growth of the ICT sector. All our case-study cities seek to be (or become) a principal centre of ICT, as can be read in the official strategy documents. The previous section has revealed general factors and circumstances that are conducive to ICT cluster development. The subject of this section is more policy-oriented. From our fieldwork in five cities, we show examples of different types of ICT cluster policy, and critically asses them.

Evidently, a number of relevant variables that determine the development of a cluster cannot be influenced directly by urban policy makers but are, rather, national or European competencies. Examples are telecom market regulation, competition policy, tax policies and research and education policies. Furthermore, the scope and room for urban policy varies widely among countries; for instance, Finland and The Netherlands are much more decentralised than France, which implies more degrees of freedom for urban government. The competencies regarding ICT cluster policies vary accordingly.

For the purpose of this study, it is useful to discern the following types of cluster-specific policies: infrastructure policies, ICT adoption policies, inward investment policies, the creation of buildings dedicated to ICT companies, and finally, the development of experimental locations with broadband connections (see Figure 3.2).

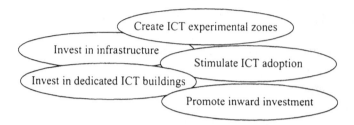

Figure 3.2   Policy instruments to promote the urban ICT cluster

Here, we have deliberately left out the internal ICT policies of the cities (such as introducing local e-government, e-democracy and service provision): these topics will be discussed in the next chapter.

Infrastructure Policies

Traditional telecom infrastructure provision (construction and exploitation) used to be a task of state-owned telecom monopolists, which owned and maintained the networks and also provided services. They provided national access without much regional discrimination. Additionally, in many countries local governments owned cable networks that transmitted television and radio programmes in the cities. In the last 10 years the situation in Europe has changed (and is still changing) drastically. Most old telecom monopolists in Europe are privatised and have to compete with new carriers and allow service providers on their networks. New infrastructure providers are developing electronic infrastructure (glass-fibre networks), some Europe-wide (KPN Quest), some national and some global (MCI-Worldcom). These firms do tend to be spatially discriminating: generally, they follow demand, which means that big 'telecom spending' agglomerations end up much better connected and served. Additionally, many of the local cable companies are 'eaten up' by large telecom providers. The former cable company of The Hague, Casema, for instance, is now owned by France Telecom. As a result, cities have lost control over their cable networks. So, what are urban managers doing to influence the electronic infrastructure in their cities? We found several strategies.

- *Connection to national/international backbones.* The city of The Hague may serve as an example. This city is not a hub on the carriers' backbone, and is therefore less accessible than, for instance, Rotterdam and Amsterdam. To change this, the city seeks to convince carriers to extend their backbones to the city of The Hague. It has commissioned an inquiry into the telecom demand potential. The Hague proved to have potential comparable to Rotterdam, but is not connected. The Hague also introduces carriers to potential clients in the region. The effectiveness of such policies can be questioned: whether carriers know the market less well than urban management remains questionable.
- *Investment in broadband infrastructure in (parts of) the city as a means to speed up the development and adoption of new services, or help specific groups.* The city of Eindhoven is creating an electronic 'freeway', free access to broadband, for some 80,000 of its inhabitants: this 'knowledge neighbourhood' is to become a test-bed for new services. National government and companies are partners in the project. Helsinki has a similar project in the Arbianranta area.
- *Creation of new infrastructure for intra-municipality data traffic.* The city of Marseilles has infrastructure rolled out in the metro tunnels, connecting several local government agencies. A critical note could be why cities invest in dedicated infrastructure for the internal data traffic, instead of leasing lines from telecom operators whose core business is the provision of telecom infrastructure.

Dedicated Buildings

Several European cities invest in dedicated buildings for ICT companies, in the hope that the concentration of ICT firms in one building will lead to synergies. A very special concept was developed in the inner city of Helsinki: the Lasipalatsi project. This is a public building that hosts the surprising combination of ICT firms, media firms and leisure activities (a cinema and various bars). The centre not only offers office space for ICT and new media firms, it also serves as 'showcase and disseminator' of new information technology developments. For instance, the bars and restaurants in the centre have touch screens built in to the tables, enabling visitors to order their drinks or to surf the Net. Also, it brings producers and consumers of IT closer to each other: generally, developers of technology, the developers of its content and the everyday user are not in contact with each other (Lasipalatsin Mediakeskus Ltd, 1999). Lasipalatsi has a social goal as well: its open character should help to promote more equality in the information society by making IT easily accessible to all. The social function justifies public investment in the project. The building is owned by the city, but exploited by a separate public/private organisation that has substantial management freedom. This arm's length construction works very well: it permits faster decision-making. The Lasipalatsi contributes to the liveliness of the inner city of Helsinki, as it attracts many visitors.

Public support is not a necessary condition for creating dedicated ICT buildings: in The Hague, a private developer has turned an old switchboard installation building (with the advantage of good external data connections) into an ICT centre. It hosts 33 small firms (total employment: 200), among which are not only software developers, Internet designers, Internet access-providers, but also non-technical ICT-related firms, such as lawyers specialising in intellectual property issues, the publisher of an IT magazine and IT consultants. This developer has, from the beginning, aimed to realise synergies and cooperation in the building. Therefore, tenants are screened for their value-added for the concept. A facilities provider in the building, Hobbit Facilities, plays a central 'bonding' role in the concept: to stimulate cooperation, it organises theme sessions, sports events, and, for instance, an organised tour to CEBIT, the main computer fair in Europe. The concept seems to work well: there is a waiting list for tenants and the number of bankruptcies is very low compared to the national average. The role of the city of The Hague has been limited: it has only offered rent subsidies for firms during the first year the building is in operation.

Experimental Areas or Neighbourhoods

Some European cities explore the possibilities of the future by building futuristic urban quarters 'loaded' with the latest technology. In that way, they try to make their city attractive as a seedbed for new technological and infrastructure development and to increase the number and quality of interactive services that are offered. Helsinki's Arabianranta area is an outstanding example, located 6 km from the centre of Helsinki. The scheme's aim is to turn Arabianranta into one of the leading ICT areas in Europe, where the functions of living, working, studying and recreation are mixed. The project is planned to be completed by 2014. The area already hosts 4,000 residents and 3,000 jobs; these numbers are intended to increase to 12,000 jobs and 8,000 residents. An integral development strategy has been adopted, in which there is space for such complementary activities as research, business, education, housing and shopping. Already, 50 to 60 firms are present on the site, employing some 300 people. The number is growing, for the site is very popular. Many of the small firms work for larger firms in the region, such as Ericsson, Nokia, Siemens and Sonera. The University of Art and Design has also been located on the site since 1985. Furthermore, new high quality housing and shopping facilities are being constructed. The electronic infrastructure of the Art and Design city is very sophisticated. A region-wide broadband fibre network is being implemented, with open access to all residents, schools and businesses. The intranet offers opportunities for teleworking, -shopping and -banking, tele-health and library services. The site proves attractive for large firms such as the Sonera telephone corporation, and Nokia. For one thing, these firms can benefit from the superior quality of the electronic infrastructure by using the area as a test site for new products, or as a showcase to display what is possible in the future.

Eindhoven is about to launch a similar project. Unlike Helsinki, the city will equip an existing urban quarter with broadband infrastructure. The quarter is very mixed, with high and low income residents, old and new houses, and stretches beyond the municipal borders. It will be an experimental zone for new services: not only will private companies offer services, but health care facilities and online municipal services will also be offered.

ICT Adoption

Many cities run programmes to enhance ICT adoption by firms in their city. In this way, they not only seek to improve the competitiveness of 'old economy' firms (mostly small and medium-sized enterprises (SMEs)), but also indirectly stimulate the ICT sector by extending its client base. Manchester and The Hague run projects to help SMEs make the shift to the information economy, by offering them free advice in the form of consultancy days (The Hague), or cheap advice (Manchester). Manchester uses ERDF money to fund these policies and leaves the technical support to a university-based organisation; in The Hague the project is carried out by Syntens, a regionally active organisation of the National Ministry of Economic Affairs. The impact of these programmes on the development of the ICT clusters in the respective cities is unclear, although we feel that the effects should not be overestimated.

Another policy option is to stimulate ICT adoption by the population. In all our case cities, local government invests in the computer skills of the population by equipping schools and public buildings with computers, or by offering training. The use of new technologies can also be stimulated in a leisure-like setting: in Helsinki, visitors to a café or restaurant in the Lasipalatsi (Glass Palace) building are directly confronted with ICT. A touch screen is integrated into the table, with which drinks can be ordered. More on this issue can be found in the next chapter. In the end, the more eager urban population is to use new technologies and applications, the larger the market for local ICT entrepreneurs.

Inward Investment

Many cities try to expand their local ICT sector with marketing efforts directed towards (foreign) ICT companies. In Manchester, the inward investment agency has designated ICT in general as a spearhead sector. It offers US- and Canada-based firms general information on the region and supports them to find a good location and staff, as well as offering tax reduction and subsidy facilities. In The Hague, acquisition efforts are tailored to telecom companies and e-commerce companies. The focus strategy pays: the city has managed to attract Amazon.com and Mapquest's European headquarters, and two large telecom providers. Marseilles wants to be an attractive location for telecom operators as well. To attract as many operators as possible, a one-stop shop agency has been created to welcome and receive new entrants. The city already hosts some 20 operators (Ville de Marseille, 2000). If cities manage to increase the number of operators, this may be beneficial for the citizens and companies when competition leads to more choice, better service and lower prices.

The regional scope of acquisition efforts varies among the cities. The inward investment agency of Helsinki, the Helsinki Metropolitan Development Corporation Oy HMDC, promotes international business not only in the city of Helsinki but in the entire region. HMDC is a joint-stock company, 52 per cent of which is owned by the city of Helsinki. Other shareholders are the municipalities around Helsinki, the Uusimaa Regional Council and the Chambers of Commerce in Helsinki and Espoo. It markets the area as a suitable location for international business and cooperation and promotes it as a major business centre of the 'new northern Europe'. IT is one of its target sectors. Also, it helps SMEs in the region to expand their business abroad. In Manchester, the acquisition organisation works on a regional basis: it is owned not only by the city of Manchester, but also by the neighbouring cities of Salford, Trafford, and Tameside. It uses 'Manchester' as brand name to sell the region. In The Hague region, on the contrary, cooperation is not so smooth: the neighbouring cities of The Hague, Delft and Zoetermeer have their own inward investment policies and strategies. The lack of coordination leads to unfruitful and inefficient inter-municipal competition. Distrust among the regional policy makers is the main cause.

6 Strategic Lessons for Cities

In this chapter we have elaborated on the role of urban regions as attractive locations for ICT companies. Also, we described a number of policy options for cities to strengthen their position in this respect. What can cities learn from these analyses? A first, general lesson could be that it is wise to give room to private initiative. To create favourable conditions for highly dynamic ICT companies to flourish is of utmost importance. This includes provision ot (electronic) infrastructure and appropriate business locations that suit today's demand, and avoidance of bureaucracy.

To speed up ICT business development, urban management can stimulate and/or enable experimentation with new technologies. Public investment in broadband infrastructure – for instance in experimental zones – might give a city a lead in ICT adoption by the population, and also speed up the development of useful services – if possible by local companies – as there it provides critical mass. This might give local companies a competitive lead in the future.

To arrive at effective policies, it is a prerequisite that urban managers should know the ICT sector very well. In too many cases policies are ill-founded, as urban officers have insufficient knowledge of what drives the market. One fruitful way to stay in touch with market developments in the ICT sector is to visit one of the many local 'First Tuesday' events, meetings of Internet firms which are held in every large city.

Public/private cooperation is a prerequisite for arriving at effective and efficient cluster policies. This holds in the marketing of the cluster, the attraction of new firms, in helping start-ups and in all other aspects of cluster policies. Public private cooperation helps to make optimum use of the knowledge and resources of the existing actors in the cluster. This again implies that officers involved in cluster policies need to be well educated and have sufficient 'feeling' for the cluster.

Besides these general lessons, more specific ones can be derived.

- *Starters' policy*. Create a starters-friendly environment: in the ICT sector, the starter of today can be the multinational of tomorrow. Local government can act as provider of generic support, such as (cheap) accommodation and venture capital, but also as purchaser of the products of new companies, or as broker and connecting factor between starters and both the existing private sector and the knowledge institutions. Incubating is a lot cheaper – and often more effective – than attracting foreign business;

- *Inward investment.* If you want to attract high tech companies. define your target well. Rather than attract just any type of ICT companies, make well-considered efforts to attract ICT companies that fit the local economic structure or that fill gaps in the existing cluster. It is unproductive to address the ICT sector as a whole, given the variety of firms' characteristics and needs within this sector. The concentration of The Hague on telecom firms shows how effective a focus strategy can be. For inward investment agencies, it pays to use contacts and networks of existing firms in the regions with others. Involvement of existing private ICT firms in the inward investment stimulation schemes improves efficiency and effectiveness.

- *ICT adoption.* Stimulate adoption of ICTs by 'old economy' firms. Under some conditions, local government could encourage or speed up ICT adoption by lagging SMEs. Instruments are education schemes, demonstration projects, the provision of ICT consultancy days at reduced fees. Moreover, the city can organise working group sessions where ICT companies meet 'old economy' firms in the region.

Chapter Four

ICT as Instrument to Promote Urban Attractiveness

1   Introduction

The previous chapter focused on the attractiveness of urban regions for ICT companies: we identified key location factors and showed how several European cities try to boost their ICT sector. We also derived policy lessons for urban regions to help them to profit from the growth of the ICT sector.

This chapter elaborates on urban attractiveness in a broader sense. In our conceptual framework (see Chapter Two) we pointed out that urban attractiveness can be defined as the degree to which welfare elements can be accessed by population, companies and visitors. Welfare elements for the population include the availability and accessibility of shops, leisure facilities, job opportunities, health care, quality housing and safety. Welfare elements for firms include the availability and accessibility of appropriately skilled labour, inputs – this can be not only physical inputs, such as components, but also nonmaterial inputs such as business services or partner firms – and good business locations. For tourists or visitors, what counts is the availability and accessibility of tourist attractions, events, cultural facilities, hotels, etc.

In this chapter, we will illustrate in more detail how ICT solutions can improve the accessibility of welfare elements with examples from several European cities. We focus on four topics. First, we consider how ICT can improve the internal accessibility of an urban region transportation (section 2) and contribute to the quality of public services (section 3). In section 4, we discuss the importance of Internet access for the entire population and show how European cities deal with that challenge.

For each topic, we display the technological opportunities and discuss organisational issues, illustrated by examples from European cities. Note that this chapter is far from complete: it provides only a few examples of how cities might deploy ICT. The chapter ends with strategic policy recommendations.

2   General Accessibility: Quality of Urban Transport

Many European cities suffer severe problems of congestion. ICT is an instrument which can improve accessibility in urban regions in several ways. Here we confine ourselves to two, namely ICT as a tool to: 1) fight road congestion; and 2) improve public transport services.

First, ICT can help to manage road traffic flows. Improved information on the traffic situation – Internet sites showing the traffic situation, or all kinds of information signboards on highways – is a great help for citizens and visitors, enabling them to make a better informed choice and a more efficient use of existing infrastructure. More possibilities emerge with the introduction of mobile Internet services and when cars become more intelligent, i.e. are equipped with electronics able to receive traffic information and translate it into alternative routings if necessary. More intelligent systems may propose new routes in case of queues or delays.

In Marseilles, to improve the information provision on the traffic situation and the possibilities of public transport, a partnership of metropolitan actors is now developing a central server that should process this information and make it available to citizens via the Internet and Minitel. The aim is to promote the use of public transport and ensure a more efficient use of the road network in the agglomeration. Principal partners in the project are the transport companies in the region, SNCF (the national train operator) and RTM (Marseilles public transport), operators of the road network and the city of Marseilles. Each partner is to provide information to the central server named 'Lepilote', which is managed by an organisation with the same name. Figure 2.1 below shows the providers of content, the management of information and the distribution to end users.

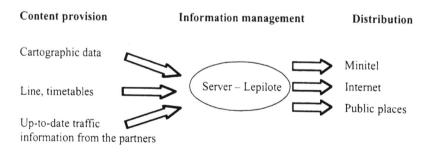

**Figure 4.1   Information: content delivery, information management
                  and distribution**

ICT offers possibilities not only for improving information provision to travellers, but also improving the connection of different types of public transport (bus, metro, trams, railways). Too often in public transport, the provision of information is supply- instead of customer-oriented. ICT may help to improve this. An effective way is to create a 'virtual front office' that presents all the public transport operators as if it were one company. One way to achieve cooperation is to introduce smart cards that can be used for all kinds of public transport.

To improve interconnectivity in its public transport system, in the 1990s Marseilles introduced a magnetic card that permitted payment for both parking at the transfer points and the different types of public transport. A central information system registers the sales of cards at the (numerous) selling points, as well as the use of the cards in the various transport companies, and redistributes the revenues accordingly among the participating firms. The introduction of the system was a difficult exercise, as many players had to be persuaded to participate. On the track from Aix-en-Provence to Marseilles alone, some 18 bus companies had to be brought together to cooperate. Recently, the city transport department decided to develop a new card with a microprocessor that would permit more information storage, allow more personalised and customised functions, and be less susceptible to fraud.

Critical Issues

Technically, the opportunities to deploy ICT in transport are endless: we provide only a few of many examples. In reality, it proves hard to implement even relatively simple solutions. Why?

Often when a new information technology is introduced, a critical success factor is the number and type of partners who join. Positive network effects are at work, as the chances of success of a system are the greater the more actors cooperate. The introduction of the magnetic card in Marseilles, for instance, was so successful because so many actors decided to participate. In many cases it is very difficult to convince all partners of the need for a new system.

In some cases, strong leadership is needed to achieve success. A 'network champion' who pushes the new technology and takes the lead in its development and implementation can do much to speed up adoption by other actors. In Marseilles, the strong position of the municipal transport department (RTM) greatly contributed to the adoption of the magnetic card. In other cities, leadership might be less strong. It follows that urban organising capacity –

the ability to enlist and involve all relevant actors – is a critical capability of the municipal organisation in the information age.

3   Quality of Public Services: E-government

For some years already European cities have tried to make their administration more responsive and improve the quality of their services, as a weapon in the competition with other cities to attract firms, people and visitors. For instance, many cities have opened one-stop shopping agencies for specific target groups – investors, property developers, or even film makers! – to ease their way through the urban bureaucracy.

The possibilities of the Internet greatly enlarge the cities' possibilities in these respects. *The Economist* (2000) calls it the next revolution, after e-commerce. It states that 'reinventing government ... is at last being made possible by the Internet', with all the beneficial impact for the business climate and citizens. Indeed, Internet technology makes it possible to substantially improve the service provision of local governments.

Table 4.1 shows how local governments can use the Internet. In the most simple case, each municipal department puts some information on the web. In a somewhat more sophisticated case, each department can add interactive or even transaction features. These models, however, are still supply-oriented: the final client (citizen, tourist or company) is not interested in the departmental make-up of the municipality but in answers to his questions. Needless to say, this supply-based form is far from optimal for the customer, who has to surf through the jungle of municipal websites to find the right service, if it is available at all.

The right-hand side of the table shows the client-oriented approach: here, the city has a virtual front office where all municipal services are offered. A system can be designed to process all kinds of information and interactive services to citizens and business, 24 hours a day. Eventually, systems can be designed tailored to the personal needs and circumstance of the user: you only get the information and services that you need.

Practice in European Cities

Most European cities are not that advanced, although things are changing fast. Most cities have no single front office for the local government services: at best, each separate department has its own site. Furthermore, the interactive

Table 4.1   Public services on the Internet: stages in e-services

Orientation / Function of the site	Supply-oriented (fragmented)	Client-oriented (integrated)
Information provision	•	•
Interactive features	•	•
Transaction features	•	•

features of the websites are very restricted. Marseilles' municipal website is an example of a demand-oriented service. It addresses three main groups: visitors, inhabitants and companies, for which it organises the information in different ways. With a search engine visitors can find the right municipal department for specific issues and questions. Citizens can order official notifications (on marital status and date/place of birth) directly by the Internet, by filling in an online form. More interactive features are to be added (see http://www.mairie-marseille.fr/). Marseilles wants to go a step further and aims to add other services as well (examples are online health services and information on traffic). The city has developed a general interface by which different servers can be linked functionally, while at the same time staying independent. The interface is to be completed with an intelligent search engine that should turn the web into a very customer-friendly product. This search engine should be able to provide answers to multifaceted questions: for instance, if you look for a cinema show in the city, you would get not only all the movies available at that time, but also information on tickets, online ordering possibilities, suggestions for (public) transport connections to the cinema including real-time information on delays, etc. This can be considered a next step in a client model.

The city of The Hague uses a similar model, but links it to a project aimed at Internet adoption by the population. The city has created an 'urban portal' named resident.net, in cooperation with Casema (a cable operator) and KPN (a telecom provider). On the website you enter the virtual city of The Hague, where not only local government information but many other features are offered: the portal includes squares (where people can chat); a link to the local newspaper, tourist information and more. Every inhabitant of the city gets free access to a limited part of Internet. Also, information can be retrieved via interactive teletext; with a telephone and a TV, the Internet can be browsed without a PC. The project also aims to make the many different websites of the city more accessible, by constructing a search engine and creating improved directory structures. The city of The Hague has invested EURO 1.13 million in the project. In Eindhoven, local government experiments with e-service provision for a small group: the students and staff at the university campus. All the students have a notebook at their disposal and the campus is equipped with broadband infrastructure. For Eindhoven's urban management, this makes it the ultimate test location for its new services.

E-government does not stop at e-services. Electronic media can also be used to organise consultation, for instance for large urban development plans. This not only strengthens democratic decision-making, it may also improve the quality of projects. The city of The Hague is now experimenting with interactive policy and decision-making, in cooperation with the University of Leiden. Manchester has organised a web-based application to receive reactions – from citizens, companies and other organisations – to its new regional development plan. To maximise the impact, the city has put much effort into a campaign to market the website.

Critical Issues

Urban governments throughout Europe find it hard to shift towards e-government. It is difficult to develop and implement ICT solutions that are truly geared to the needs of the customer. We have identified the following issues and challenges:

• *Internal organisation.* E-government requires drastic organisational changes within the municipal organisation: the introduction of a single 'front office' requires smooth cooperation in the back office (the different government agencies), or even a total reorganisation. For instance, when a citizen poses a complicated question by e-mail that requires the know-how of two municipal departments – say, the real-estate department and the local tax department – the departments should know how to answer it together, in time. A complicating factor is the sheer number of procedures that municipalities use: in the case of Barcelona, over 2,000 different situations concerning procedures had to be described and defined. The procedures are controlled by many different units within the municipality. To implement an integrated front office implies a reshuffle of information flows, entailing the shifting of power relations – and subsequent power struggles – within the municipal organisation.

- *Leadership.* As long as municipal bodies are involved, the mayor (or aldermen) can use his/her power and influence to enforce cooperation. When other, external parties are involved as well (such as transport companies, tourist attractions) things get complicated. A critical issue is how to involve them.
- *Relation between public and private sector.* Electronic service provision blurs the frontiers between public and private domains. Who is responsible for the information that is offered at the 'one-stop shop'? This is a complicated matter if the services combine and deploy data that are collected, stored and managed by different (public and private) organisations. Ownership of information and liability are treated very differently by the law for public and private actors. If a citizen or company is wrongly informed, what law holds? Who is responsible?
- *Identification, security and privacy.* Many municipal services require identification. To provide these services online, safe identification procedures need to be in place. This is also a complicated matter. This holds even more for online payment: online payment procedures are in full development, but few people trust them.
- *Access.* Development of e-government is problematic since currently not every citizen has access to Internet. This might conflict with the universal service provision obligation for local governments. Marseilles regards the Internet as only one electronic medium for information provision: it uses Minitel and information signboards at public places as well. The Hague uses interactive television technology to reach all citizens with a TV as well; also, the city acts as free Internet provider and operates an intranet for all The Hague's inhabitants.
- *Relations with industry partners.* To develop, operate and maintain e-services, the technical know-how and experience of the private sector – ICT consultants or service providers – are needed. A critical issue is how urban management deals with these private partners. The risk is great that private companies, with their head start in technical competence, take the lead in systems development or take too big a slice of the cake. In the resident.net project, the city of The Hague, although it is the principal funder and initiator, has lost too much control over its project to KPN, the former Dutch National telephone company that runs the operation. Resident.net is developed as a portal of KPN, will who probably gain most from the project: not only will it make money on the helpdesk (which is relatively expensive) and the telephone costs of Internet use (KPN is the main telephone supplier in the region), users of resident.net need also to

supply a lot of personal information to KPN. Another risk is that a city becomes 'locked in' to the chosen system, which puts the city in a weak bargaining position vis-à-vis the technology supplier. A good way to select a technology partner is through a competitive bidding process for the system development and its maintenance, for a certain period of time and with clear performance requirements.

- *Who should run the front office?* Should it be the city (because it offers so many of the services involved and produces most of the information)? One disadvantage is the bureaucracy and lack of responsiveness to new technological developments. Also, the high costs are carried by the tax payers ultimately. Or should it be run by a private actor, such as a telecom operator (because they have the technological competencies)? That would require very strict control on the side of the municipality, to safeguard the quality. Perhaps better, a public/private setting could be developed to safeguard the quality and integrality of the services on the one hand and permit flexible and fast decision-making on the other. Much depends on the revenues. If the system is a success (many people using it), it could finance itself from advertisement revenues.

4   Access to Internet

In Europe, only a minority of the population has a computer; an even smaller share has an Internet connection at home. From a social point of view this means that many people do not benefit from the possibilities. This may be due to a lack of resources, but also because a number of people have difficulties in handling a computer. From an economic point of view, low levels of Internet access (and use) hamper the development of new services. For companies and governments, a critical mass of Internet users is needed to introduce new digital services in a meaningful way.

Urban policy makers (but also national governments) realise the importance of widespread Internet access, as a means for the population to access new services and become part of the information society, but also to learn skills that are needed in the labour market. Our case-study cities seek to improve Internet access and use in a number of ways. The most common way is to provide public buildings – such as libraries – with computer equipment, and enable visitors to access the Internet. In some cases this is free – in Helsinki, for instance – and in other cases a very low fee is charged. The city of Eindhoven will connect an entire neighbourhood – around 80,000 people – to the Internet (broadband connection), in its e-city project. It is hoped that many companies will offer innovative services to this critical mass of users and that citizens will extensively use the new opportunities. The city of The Hague even offers access to the entire population of the region. In the resident.net project, people can get free Internet access, but also access the Net via their television. In this way, Internet access is decoupled from PC ownership.

Some cities have designed policies to include less favoured groups in the information society. Examples are ICT education projects, or the creation of Internet access points in deprived neighbourhoods. In Manchester, the Electronic Village Halls (EVH) initiative offers access to Internet and information technology education and training for weaker social groups. The EVHs (started in 1992) offer Internet terminals where the Internet can be accessed free; moreover, numerous training and education programmes are offered at low fees. The EVH initiative is supported by the European Social Funds and ERDF, as well as the city council.

5   Conclusions and Lessons for Cities

This chapter shows how the use of ICT can enhance a city's attractiveness in a number of respects: it can improve the quality of municipal services to citizens; it may help to fight congestion, improve the use of existing infrastructure and thus make the city more accessible. More in general, it helps to increase the quality of the urban product itself, as well as its visibility and accessibility.

The technological possibilities are great. This raises the question why, then, are so few of the possibilities actually successfully exploited by European cities? We have found that organisation, not technology, is the critical issue. The main problems that arise when inter-organisational ICT is introduced are how to make the relevant partners join, how to select the right standard of technology, how to distribute costs and benefits of a new system and how to enlist industry partners.

A related issue concerns agreement of standards. The introduction of new information technology often requires agreement on standards. To get a standard accepted (which can be a smart card of a certain specification, or the architecture and layout of the municipal Internet site, or a platform for integrated services), adoption of the standard by a few leading players can convince smaller ones to join the bandwagon (Shapiro and Varian, 1998). If important players do not accept the new system – which can happen when they face high switching costs, or are not convinced of the superiority of the new technology – it is doomed to fail.

Innovation often thrives on the efforts of a few people. The website case shows that a few enthusiastic and visionary people in a strategic position in the municipal organisation can be decisive in a city's move towards interactive public service delivery to citizens and to e-government. Marseilles would not have been elected best Internet site without such people. The Lasipalatsi project in Helsinki also thrives on the energy of a few people only.

ICT can be a tool to stimulate cooperation between actors not used to cooperation. The 18 bus operators on the Marseilles–Aix-en-Provence track only now cooperate because they share the new information system. In a similar vein, within the municipality, the adoption of the Internet as the central medium for information provision and service delivery can be a trigger for intra-municipal cooperation and improve the general level of service provision even further.

The Internet offers the possibility of combining information from various sources in real time, with enormous potential added value for users. Examples are the provision of real-time traffic information on a website. Again, to reap the fruits of the new technology, (municipal) organising capacity is crucial. In this case, the central ability is how to convince all the different (public and private) content providers to bundle and regroup their information flows at one point or for one target group. The Lepilote project in Marseilles is a courageous effort to bundle information (both permanent and real-time) on road traffic, public transport and all kinds of tourist information from different sources, to create one product that is extremely useful for the customer/end user.

In the short and medium term, the role of Internet as a medium for information provision and service delivery should not be overestimated, because Internet penetration among the population is still limited. Also, in many instances, the PC is not the appropriate medium to receive information, for instance when you are on the road. Additional media should be used for public information provision. Marseilles' use of signboards at transport nodes and information pillars is a good example of information distribution by other media.

The following strategic lessons can be derived:

- *Invest in capabilities to manage technology projects.* The cases show that the implementation of (interorganisational) ICT is fraught with difficulties: how to select the right technology, how to deal with internal power struggles, how to involve the private sector, etc. Cities need highly qualified people who understand technology and have a view on urban development at the same time.

- *Create responsive organisations to execute technology projects.* In many cases, the management of technology projects can best be organised at some distance from the local government. The experience of Lasipalatsi shows that the arm's length organisational form is appropriate to manage ICT projects: it provides for the much-needed speed of action and enables new developments to be dealt with in a very flexible way. In this fast moving and ever-changing field in particular, freedom of action and flexibility are preconditions to make a project such as the Lasipalatsi a success. This implies that a different urban management style needs to be developed. Urban managers need to become more aware of technology development, but also of the needs of the businesses in general. They should be able to engage in networks with private partners and make use of their knowledge and resources, but at the same keep the general overview of the public interest.
- *Carefully debate, define and control the degrees of access to information for different users.* Civil servants need access to different information than do citizens or companies. Marseilles' solution, with an intranet for internal use, and an Internet site targeted at different user groups (citizens, visitors and companies), is clear, and seems to works reasonably well.
- *Never give the lead to the technology suppliers.* To make technology an instrument instead of a goal in itself, urban managers should first and foremost have clear ideas on what they want from the new system, and under what conditions, and then let the supplier come up with appropriate solutions. To avoid being 'locked in' by a single telecom operator or software house, cities should be very wary of binding themselves to a specific system designed by a single supplier. They should be careful in contracting and ensure escapes to other suppliers in case of exceptional price charges or under-performance. To organise competitive bids, including service and maintenance contracts, can be a means to elect the best supplier.

Chapter Five

Outlook and Further Research

1   Introduction

A broad range of ICT policies has been discussed. In this final chapter of this part of the book, we will provide some concluding remarks, notably on how to integrate ICT projects into urban development policies. Finally, we suggest issues that need further research.

2   Guidelines for Urban ICT Policy

The information revolution changes urban life in many ways. However, despite its pervasive power, ICT is not a force that descends upon urban societies and changes them in a deterministic way: it is rather a complex of technological possibilities that can be used, shaped and directed by urban managers and politicians. They can use the new technology to improve the delivery of municipal services to inhabitants and businesses, or to increase civil participation and local democracy by erecting electronic discussions platforms; they may help their citizens and businesses to make the shift to the information society by offering them free Internet access or education facilities at low prices; they may design specific policies to boost the local ICT sector or attract companies from outside. They may create experimental zones or neighbour-hoods where new services and products can be developed and tested. In sum, ICT offers opportunities to strengthen a city's economic and social profile.

To grasp the opportunities, new institutional arrangements are needed. The most important job for policy makers is to create an institutional environment that supports technological change.

In our view, a principal role for urban government is to prepare its citizens for the digital era. This can be done by offering Internet access and education, particularly to the weaker social groups. In the design and implementation of urban ICT projects and policies, a bottom-up approach is advisable. Urban management should give local players – schools, social organisations, churches, health care institutions, firms, etc. – the opportunity to come up with innovative ICT projects and reward good proposals. Other funds – national or European – could be channelled to such projects as well: this would reduce the current fragmentation of ICT policies at different administrative levels. A fruitful way to improve the quality of projects is to link local projects to panels of professionals with expertise in the relevant field; in other words, to 'twin' experts to local projects. The advantage of this approach is that enthusiastic people within organisations – who are crucial in renewal processes – get more room to realise good projects. There are no blueprints for ICT projects yet: therefore, close consideration should be given to experimentation, learning and knowledge exchange.

Before launching urban ICT projects – in whatever fields – benchmarking can be a good way to start. Experiences in other cities can yield very valuable information on opportunities and pitfalls of ICT projects. It is particularly advisable to invest in international comparisons and translate foreign experiences where and when possible to the national and local context.

3   For Further Research ...

This brings us to themes that need further research. This study was predominantly explorative and broad in range: in a very general sense, we studied how cities deal with ICT. We have touched upon many issues that require elaboration. It is our conviction that in a number of urban policy fields comparative studies can yield important lessons. To list a few: ICT and urban transportation, ICT and health care, ICT and safety, ICT and social inclusion and ICT and city marketing. In each case, best practices could be identified and analysed, yielding lessons for cities that want to run similar projects. As the topics are so new, benchmarking is a good means for cities to 'cherry pick' and avoid mistakes.

A shared issue concerns how public, semi-public and private organisations can successfully cooperate in new ways. This is a pressing issue, as organisational frontiers are blurred with the application of interorganisational ICT projects. For urban managers, it is a major challenge to find the right public/private cooperation models. It would be interesting to analyse good practice constructions for complex interorganisational ICT projects in Europe, and discover the conditions that should be met.

A particularly interesting, complex and pressing topic is how to create responsive 'virtual town halls'. How can cities improve service provision using the Internet? An in-depth comparative analysis of European cities could reveal which factors are relevant, what pitfalls should be avoided, what organisational changes are needed, and so on. It would be interesting to include 'leading' cities in this analysis, maybe also US and Asian cities. A related question is how the relationship between government and citizens can be improved with e-democracy instruments.

References

Abler, R. (1977), 'The Telephone and the Evolution of the American Metropolitan System', in De Sola Pool, I. (ed.), *The Social Impact of Telephone*, MIT Press, London, pp. 318–41.

Alles, P., Esparza, A. and Lucas, S. (1994), 'Telecommunications and the Large City–Small City Divide: Evidence from Indiana cities', *Professional Geographer*, No. 46, pp. 307–16.

Berg, L. van den (1987), *Urban Systems in a Dynamic Society*, Gower, Aldershot.

Berg, L. van den, Braun, E. and Meer, J. van der (1997), *Metropolitan Organising Capacity*, Ashgate, Aldershot.

Berg, L. van den, Braun, E., Meer, J. van der and Otgaar, H.J. (2000), *Inner-cities of the Twenty-first Century*, Euricur, Erasmus University, Rotterdam.

Berg. L. van den. Braun, E. and Winden. W. van (2001), *Growth Clusters in European Metropolitan Cities: A new policy perspective*, Euricur, Erasmus University, Rotterdam.

Booz, Allen & Hamilton (2000), *The Competitiveness of Europe's ICT Markets*, Ministerie van Economische Zaken, Ministerial Conference, 9–10 March, Noordwijk.

Bramezza, I. (1996), *The Competitiveness of the European City and the Role of Urban Management in Improving the City's Performance*, Tinbergen Institute Research Series, No. 109, Erasmus University, Rotterdam.

Braun, E. and Meer, J. van der (2000), *The Dublin Challenge: Managing growth*, Euricur, Erasmus University, Rotterdam.

Campus Ventures News (1998), Issue 4, Statistics, p. 3.

Centraal Bureau voor de Statistiek (CBS) (1999), *ICT-markt in Nederland 1995–1998*, Voorburg/Heerlen.

Financial Times (2000), European Venture Capital Report, 7 June.

Financial Times (2000), 31 July, by Nikki Tait in Chicago and Tom Foremski in San Francisco

Graham, S. and Marvin, S. (1996), *Telecommunications and the City: Electronic spaces, urban places*, Routledge, New York.

Hall, P. (1995), 'The Future of Planning', in Giersch, E. (ed.), (1995), *Urban Agglomeration and Economic Growth*, Springer, Berlin.

Hall, P. (1998), *Cities in Civilization*, Pantheon Books, New York.

Kunzmann, K. (1996), 'Euro-megalopolis or Themepark Europe?', *International Planning Studies*, vol. 1.

Lasipalatsin Mediakeskus Ltd (1999), *Annual Report Urban Pilot Lasipalatsi 1999*.

MIDAS (2000), *Environment for ICT*, internal publication.

Mitchell, W.J. (1999), *E-Topia: Urban Life, Jim – But Not As We Know It*, MIT Press, Boston.

OECD (2000), *Information Technology Outlook 2000*, Paris.

Ohmae, K. (1995), *The End of the Nation State: The Rise of Regional Economies*, The Free Press, New York.

Porter, M. (1998), 'Clusters and New Economics', *Harvard Business Review* No. 6, November/December.

Rotterdams Dagblad (1998), *Nieuw glasvezelnet koppelt Rotterdam aan wereldcentra*, 17 June.

Saxenian, A. (1994), *Regional Advantage: Culture and competition in Silicon Valley and Route 128*, Harvard University Press, Cambridge, MA.

Schmand, J., Williams, F., Wilson, R. and Strover, S. (eds) (1990), *The New urban Infrastructure: Cities and telecommunications*, Praeger Publishers, New York.

Shapiro, H. and Varian, R. (1998), *Information Rules: A strategic guide to the network economy*, Harvard Business School Press, Boston.

Shields, P. Dervin, B. Richter, C. and Soller, R. (1993), 'Who Needs POTS-plus Services? A Comparison of Residential User Needs along the Rural–Urban Continuum', *Telecommunications Policy*, vol. 17, pp. 563–87.

Swijngedouw (1998), 'De Dans der Titanen en Dwergen: "Glokalisatie", Stedelijke Ontwikkeling en Groeicoalities – Het Brusselse Enigma', *Tijdschrift voor de Belgische Vereniging voor Aardrijkskundige Studies*.

The Economist (2000), Survey on Government and the Internet, 24 June.

Ville de Marseille (2000), *Marseille Economie, Bilan d'étape, Année 1999*, Direction Generale du Developpement Economique.

Vliet, F.L. van (1998), 'De change agent en zijn resources, een modelmatige benadering van regionale technologische veranderingsprocessen', dissertation, Eburon, Delft.

PART THREE
CASE STUDIES

Chapter Six

Eindhoven

1 Introduction

This case study describes and analyses the 'kenniswijk' (e-city) project, an innovative project in the city of Eindhoven that aims at speeding up the development of the information society in a designated area with a population of 84,000. In parts of Eindhoven and the neighbouring city of Helmond, homes will be connected to broadband infrastructure; companies, local government and other organisations will develop all kind of services to the population, and new virtual communities are expected to form. The project is a test-bed for the information society of the future.

At the time of writing, the project has not yet taken off. Therefore, this study cannot evaluate the pros and cons of the project. However, the project is innovative and promising, and contains a number of elements already in place. Interestingly, the Eindhoven University of Technology already has experience with a comparable project: every student has a notebook, the campus is equipped with broadband infrastructure, and new services are rapidly being developed. Thus, the Eindhoven campus can be considered a 'pilot within a pilot'.

In section 2, the background and aims of the project are sketched. Section 3 describes the several elements of the project: the way citizens are connected, how they are equipped with infrastructure, and what services will be developed. Section 4 focuses on the notebook project at the Eindhoven University of Technology. Section 5 discusses the organisation of the project and the importance of regional organising capacity as success factor in these type of projects. Section 6 concludes.

2 Background and Aim of the Project

The e-city began as an initiative of the Dutch government, notably of the Ministry of Transportation and Telecommunication. The Dutch national government has adopted an explicit strategy to speed up the transformation of the Dutch society into an information society. Several policy instruments are used, such as the development of electronic highways, an integrative support scheme for ICT starters ('twinning'), the provision of Internet access for less favoured groups, and many more. The e-city project is a recent innovative instrument.

Every Dutch city had the opportunity to come up with a plan to design a 'knowledge neighbourhood' in the city. Several conditions had to be met: the area should host a mixed population in terms of income, ethnicity and social position; the project should offer innovative services and provide adequate infrastructure and access; public and private players should be involved and private investment in the area should be substantial; (elements of) the project should have potential to be applied on a larger scale in due time. Also, the organisation of the project should be very sound.

Several Dutch cities submitted a project proposal. In July 2000, the state commission decided that Eindhoven's proposal was to be rewarded. The city's proposal was particularly praised for its impressive number of pilot projects, the quality of public/private partnerships, the balanced composition of the quarter and the sound organisational concept. The Ministry provides EURO 9 million, to support the investments in infrastructure. The cities of Eindhoven and Helmond invest another EURO 5 million, for the period of two years (*De Volkskrant*, 14 July 2000).

In its proposal, Eindhoven defined the achievement of 'better living, better working and better learning' by deploying information technologies as the principal aim of the e-city. Furthermore, the project should create conditions for accelerated time-to-market of new services, the testing of innovative ICT applications in the field of consumer services and infrastructure.

3   Elements of the E-city

The e-city's core area has a population of 84,000 (38,000 households) and covers parts of Eindhoven and Helmond. It includes residential areas of widely different character, two city centres, business parks and office locations, as well as the campus of the Eindhoven University of Technology and the polytechnic. It contains new subdivisions, but also reconstruction areas. The diversity may reveal differences in the way people deal with new technologies and new services, and may yield important lessons for companies and government. The size of the area is such that a critical mass of users with broadband access is reached: this is a very important condition to serve as a test-bed for new services. The e-city project has two main components: infrastructure/access provision and service development.

Infrastructure and Access

For an e-city, appropriate infrastructure is a 'conditio sine qua non': every household should have broadband access. Eindhoven has a favourable initial position: the ICT infrastructure in the area is already well-developed and delivered by a variety of companies and the region counts 17 telecom providers, using several types of infrastructure (cable, fixed and mobile telephone). KPN (the principal telephone operator) will speed up the enrolment of ADSL infrastructure. Mobile operator Libertel operates a GPRS pilot project in the region (the first in the Netherlands), offering 115 KB/sec. The pilot will be extended to the e-city. The same company will experiment with new value added services such as WAP and blue tooth. The cable operator, UPC, can increase its data capacity at relatively low cost. Further, it will offer a digital decoder/set-top box for television, which will permit new services such as Internet access through television.

In the e-city, minimum bandwidth should be 2MB/sec. At relatively low costs, both KPN and UPC are able to offer this in the e-city area in due time (mobile services at the speed of 2MB/sec. are not expected before 2002). Some critics argue that 2MB is hardly sufficient for the smooth operation of services. However, to raise bandwidth for all citizens to higher levels would substantially raise investments.

Each inhabitant of the e-city will have access to the services by PC, the TV or the mobile phone. Not every e-citizen has a PC. Several companies in the region offer funds or leasing opportunities, enabling citizens to obtain a PC at a relatively low price. The TV can also be used to access the services, with set-top boxes that are supplied by Philips and UPC. Finally, the mobile phone can be used: CMG (an ICT consultant/developer), is developing the technology to access the e-city services. A wireless local loop will be developed to permit wireless connection to the network with a laptop computer.

Content: Services and Projects

The area is to become a place where new services that add to the well-being of the population are developed, tested and implemented. The programme focuses on accelerating and improving service provision in the fields of services, health care, welfare and education/training. Wherever possible, use is made of already-existing e-services or initiatives. Among many others, the following services will be developed:

- *Government services.* Services for which identification is not necessary will be offered at a virtual desk by both the city of Eindhoven and Helmond; also, all kinds of government information will be accessible online.
- *Community formation.* Virtual communities will be created. People in the same block or even street can meet in cyberspace and form new communities.
- *Consumer services.* The e-citizen can shop electronically: his or her goods will be delivered at home. Several shops are participating in a pilot project. A construction company has organised an online auction pilot; social housing providers will show their files on Internet and offer a new range of services for their tenants such as complaint reporting, meal services, and insurances.
- *Health care, social services and leisure.* An insurance company is experimenting with online services; the public library is going online; a travel agency is experimenting with online booking; all cultural facilities and events are presented on the net, including booking facilities.
- *Education.* A college in the e-city is developing a website as interface between the home and the school, containing schedules and results.
(For a full list of activities in the above mentioned fields, see www. zebranet.nl).

Furthermore, a number of innovative projects –120 in total! – will be executed in the e-city. To list a few: many homes will be provided with smart features that enhance convenience and safety for special groups, such as chronically ill, elderly or disabled people; a demo-home will be opened, provided with experimental technologies and features; Philips Design is developing concepts to integrate the physical and the virtual world, using new interfaces such as object-integrated touchscreens; at several locations, Internet courses are offered almost free, to increase digital literacy.

A project worth mentioning is being developed by a school: it plans to offer online professional support for students in their self-directed projects. Students can invoke online support – using videoconferencing technology – from a remote teacher at fixed hours. The school mainly employs disabled teachers for this function. It gives these teachers a new role (which many of them greatly appreciate); it also helps to reduce the enormous shortage of teachers.

4   Pilot within a Pilot: The Campus Notebook Project

The e-city project will link many people to the Internet and generate a number of new services. The critical mass of users constitutes a market that is sufficiently large for new services to be offered.

Interestingly, a similar project was started at the Eindhoven University of Technology already back in 1996. From that year on, every first-year student at the university has been equipped with a notebook computer at favourable conditions.[1] The university campus has a broadband network. Students can plug in anywhere at the campus, and have access to Internet, e-mail services and all kinds of faculty services. Thus, the university campus can be considered as a test-bed: a critical mass of users (6,000 students, 4,000 staff) with broadband access is already in place. It is the greatest 'wired community' in Europe. In the course of the year 2000, student homes will get broadband connection as well. To make the transition to wireless, the university management has plans to erect a GPRS mast at the campus, enabling anyone to communicate with anyone for free, wirelessly.

The university management initiated the notebook project – and invested accordingly – in the expectation that once the infrastructure and access were in place, services and applications would soon be developed in the university community. Indeed, since its introduction, many educational applications have been developed by teachers (often helped by students!); also, students use their notebook to work on papers cooperatively.

Recently, an e-commerce professor took the initiative to go one step further and use the campus as a testbed for e-commerce. A project named VirTUE enables firms to offer new generation services to students and experiment with new technologies for very receptive customers. For instance, a supermarket is seeking to test intelligent agent technology to improve its e-services; a virtual shopping plaza has opened with many online shopping possibilities. Particular attention is paid to the development of safe identification and payment systems. Local government will also open a virtual desk at the campus to experiment with the newest services.

Students participate in the development of business applications, but also of new concepts for themselves. For instance, a virtual student bunker is being constructed, a cyber-community with a number of innovative elements where you can meet fellow students in 3-D.

5   The Role of Organising Capacity

A unique strength of the Eindhoven region is the cooperation between government, business and knowledge and education institutes. The Eindhoven region has been shown on previous occasions to have a high level of organising capacity. In 1998, it managed to attract a Twinning Centre, a national centre for start-ups. As soon as it became known that cities could submit a proposal to such a centre, leaders in the public and private sector jointly made a plan and raised money to co-finance it. Key people in the public and private sector know each other well, are willing to cooperate and invest in innovative projects, and trust each other.

In the e-city again, the region's organising capacity proved tobe a key success factor. More successfully than other applicants in the Netherlands, it managed to develop a substantial plan of action with many projects and participants in a very short period of time. The project is backed and supported by business leaders and the university management and strongly promoted by local government.

The plan of action has been developed by the municipalities of Eindhoven and Helmond and the Eindhoven region partnership. The organisation of an e-city is a complicated matter: many parties – public and private – have to be brought together to cooperate, initiatives have to be harmonised and new initiatives have to be developed. A strong project organisation was formed to write the bid-book, bring the partners together and create an integrative proposal that would meet the demands of the Ministry of Transport and Communication. At the time of writing, the e-city's foundation is developing a business plan, to be executed in a public/private construction. The project will run for two years.

6   Conclusions

The e-city project is a promising and innovative project, and can serve as example for other cities. The integrative approach of infrastructure provision and content development for a large – but diverse – group of people, will accelerate the development of new services for the population, and speed up their adoption. Companies, citizens and local government can gain unique experiences, with high returns in the long run.

Participating companies can build a competitive advantage by learning how to develop services and exploit them on a large scale. They may apply this knowledge later on a regional, national or international scale.

For the inhabitants of the e-city too, the advantages are great. The project greatly enlarges their possibilities and opportunities in the fields of shopping, learning, health, and also in the development of new social relationships. For local government, the e-city provides an ideal test-bed to experiment with new online government services.

The project may give the region a lead in the information age, if the e-city manages to scale up the project, or parts of it, to the entire region. This may also have a beneficial impact on the image and business climate of the region.

A major factor in the development of this project is the power of regional organising capacity. This project again shows how public and private leaders in the region share the ambition to improve Eindhoven's position, and dare to invest jointly in innovative projects. In future-oriented technology projects like e-city in particular, where typically many actors are involved and risks are relatively high, this shared ambition, mutual trust and innovative spirit are great assets.

The success of the project will critically depend on the willingness of the population of the e-city to make use of the new services and projects. This implies that the equipment and infrastructure should work at all times, and that in case of failures, immediate help can be offered. The universities' notebook experience shown that physical helpdesks are crucial in this respect.

The national government – notably the Ministry of Transport and telecommunications – has played a pivotal role in the e-city: it organised a 'beauty contest', letting Dutch cities come up with plans and rewarding the best one. For other national governments in Europe – but also for the European Commission – this can serve as a model of how to gear technology policy to regional needs using the regional networks and expertise, and to reward those cities or regions that are best able to organise themselves.

Note

1   The notebooks cost around EURO 3,500. The student contributes one-third, the university funds one-third as well. The final third is a loan that is paid off when the student graduates.

References

De Volkskrant (2000), 'Kenniswijk komt in Eindhoven en Helmond', 14 July.
Zebranet (2000), *Bid-book, Eindhoven E-city.*

Discussion Partners

Mr E.Z. Frank, City of Eindhoven, Technology Process Manager and Project Manager, E-city.
Mr W.G. Kolenbrander, One Way Wave Services BV, Managing Director.
Ms E.P.J. Lemkes-Straver, Economic Development Corporation Eindhoven Region, Project Manager, E-city.
Mr M. Rem, Eindhoven University of Technology, Rector Magnificus.
Mr F.J. Slobbe, EUTECHpark, Managing Director.
Mr A.L.H.C. v.d. Staak, Bisschop Bekkers College, Board Member.
Mr T.F.L. Veth, Centre for Electronic Business Research & Application, General Manager.
Mr R.P. Waterham, Eindhoven University of Technology, Head of Unit, ICT Services.

Chapter Seven

Helsinki

1   Introduction

New media and information technology are at the centre of interest for policy makers on several levels, because of their profound impact on people's lives and on economic developments. Until now, the move towards the information society has been predominantly technology-driven. However, there is a wide gap between the development of ICT and new media on the one hand and its use on the other. For instance, despite all its possibilities, even at the beginning of the twenty-first century only a minority of the people in Europe use the Internet. To speed up the adoption of new ICT, many policy initiatives are now directed to the integration of technology into everyday life and to making it more accessible for weaker social groups. A principal, yet unanswered, question is how this can best be achieved.

This case study might provide some insights in this respect. It centres around an innovative project in Helsinki named Lasipalatsi ('glass palace'), a media centre in the inner city of Helsinki. The concept of this public building is directed to the dissemination of innovations in ICT and new media, bringing technology closer to the people, and improving contact between technology developers and users. The combination of these functions in a leisure-like setting is new for Europe. For the city of Helsinki, the specific public/private organisational form of the project is innovative; in addition, for Helsinki the project was the first large project to make use of financial backing from the EC.

The case study is based on a literature study and a number expert interviews with a variety of key people involved in the project and in the city administration. It is organised as follows. In section 2 we describe and analyse the project itself: its history, objectives and characteristics, opportunities and threats. Section 3 focuses on the context of the project: how does it fit within developments in the ICT/media sector, and in the policies for the (inner) city of Helsinki? Section 4 concludes.

2   Lasipalatsi

The Lasipalatsi building is located in the heart of the city of Helsinki. The basic idea of the centre is to provide open access to information technology services and to integrate new technology into people's everyday life. In this section, we elaborate on several aspects of the project: its functional concept, its alternative organisational setting and financial issues. Finally, we sketch some opportunities and threats for the future, and draw conclusions.

Lasipalatsi: An Innovative Concept ...

Originally, the Lasipalatsi building was a temporary construction planned as a recreational centre for the Olympic Games of 1940. The building was completed in 1938. In the decades that followed, the building was saved from being demolished. In the early 1980s there were ideas to make it a location for public services. During the early 1990s a plan to transform it into a media centre with a public function was formulated. In March 1993, the Helsinki City Council announced its intention to renovate the Lasipalatsi building and convert it into a centre of culture, leisure and media. This plan fitted the new strategy for the inner city that was designed at that time. With the help of the European Urban Pilot project – the Lasipalatsi is one of the first projects in Finland with EU support – and the efforts of some very enthusiastic people, the plans were realised: the dilapidated building was completely renovated (cost: EURO 8.4 million), and a concept was developed to turn the centre into a media complex.

The goals of the Lasipalatsi can be discerned at several levels. From a cultural heritage point of view, the goal was to revive the old bazaar-like building and its immediate surroundings into an active part of the city centre. Second, the centre should serve as 'showcase and disseminator' of new information technology developments: technical facilities and opportunities are almost unbounded but the use of technology is lacking in social content, or at least is developing at a different pace from technology. Third, the Lasipalatsi should bring producers and consumers of IT closer to each other: generally, developers of technology, the developers of its content and the everyday user are not in contact with one another (Lasipalatsin Mediakeskus Ltd, 1999). Lasipalatsi has a social goal as well: its open character should help in promoting more equality in the information society by making IT easily accessible to everyone. A strategic goal of the project is to contribute to the transition to an information society. Importantly, it was decided that the Urban Pilot Proposal should serve as strategic guideline for the elaboration of the concept. Thus, the Lasipalatsi proposal played an important 'structuring' role in the realisation and functioning of Lasipalatsi.

The centre was officially opened in November 1998. In its first year of operation, the building was visited by an unexpectedly high number of 2.3 million inhabitants. The building proved especially popular among young people and females (Lasipalatsin Mediakeskus, 1999).

The Tenants

A fundamental premise of the Lasipalatsi concept was to have complementary companies and institutions in the building, to make the sum of the project bigger than its parts. Because so many institutes and firms were interested in the building from the beginning, a competitive bid procedure was held to select the right tenants. One criterion was that applying firms should make use of the Internet/intranet. Another was that large commercial firms should not be the only type of business to move in; tenants should also include some small, innovative firms that could be expected to make an above-average contribution to the Lasipalatsi concept.

There are more than 20 firms and institutes in Lasipalatsi. The tenants are very diverse. There are four basic categories: 1) private firms that are active in the media sector: several TV stations, a cinema-exploiter, production firms; 2) bars and restaurant facilities: a high-quality restaurant and several cafés, among which is the first Internet café in Europe; 3) non-media private firms: a bookstore; a flower shop; and 4) public services, among which are an ICT library and the Helsinki Festivals organisation. The cinema REX has a special role: during the day, it is leased out for all kinds of seminars related to the information society; after 6 p.m., it functions as a film house for alternative movies.

It was hoped and explicitly striven for that the tenants in the building would cooperate with each other and engage in joint projects. After one year, the concept has not entirely worked out as expected/wanted in this respect: notwithstanding some successful joint projects, cooperations are scarce, and several companies regard Lasipalatsi as nothing but a nice location in the inner city (see also Kanninen, 1999). In Lasipalatsi, the the Nykyaika ('Modern Times') organisation is the most active organising projects. It stimulates discussions between developers and users of ICT in many different ways; it organises all kinds of exhibitions related to the information society and workshops in which ICT developments are tested by the public. Nykyaika is funded by the national fund for research and development Sitra.

Official aims of Lasipalatsi

Innovativeness: a new media centre combining existing functions in a new way, in which the people of the city may experiment, observe, participate or just loiter.

Exemplary character: the unification of several participants, public and private interests with a view to common goals.

European partnership: the exchange of knowledge and experience between European centres in real time.

Financing: restoration, maintenance and development will be effected for the benefit of the whole in association with the involved participants.

Employment effects: in addition to the renovation and equipping of the centre, personal experience of the possibilities of technology may produce ideas for self-employment, product development and enterprise.

The city centre: a functional meeting point housed in an already-existing city building will arise alongside the historic and commercial centre of Helsinki.

The renovation of old buildings: renovation of a valuable building for the needs of new technology while respecting its original appearance. The open-plan character of the structure is admirably well-suited to this.

Conservation of the built-up environment: following the avoidance of its planned demolition, the restoration of the original appearance of the Lasipalatsi building will influence the appearance of the area and other planning of the city centre.

The use of information technology: 'Living room use' equipped with the latest technology will lead to cooperation between designers and users, which will create the conditions for innovations of content for the future.

Example of a public tenant: the cable book library

The Cable Book Library at Lasipalatsi is a branch of the Helsinki City Library, but a very special one. It serves as an information site and meeting place, and offers customers more than 20 workstations with access to Internet. They can use the worldwide web, e-mail, word processing and spreadsheet software, databases, layout, graphic and image-editing software. Customers can make an advanced reservation for a workstation. The Cable Book Library has a limited collection of books, specialised in cinema, Internet, travel, media, arts, comics and CD-ROMs, because space is limited and rent levels are very high.

However, all books from other Helsinki libraries can be ordered online at the Lasipalatsi location and are delivered if wanted. There are subscriptions to more than 100 periodicals and newspapers at the Cable Book Library. Subscription to the library is free of charge (as in all of Finland). The library is entirely publicly financed. It received 300,000 visitors in 1999, 30 per cent of them foreigners (Berndtson, interview). This high figure is partly due to the long opening hours.

Example of a private tenant: MTV3/TV
Lasipalatsi hosts several TV/production companies, both public and private. MTV3 is one of the private broadcasters/producers. The production company has 15 employees, a turnover of EURO 3.3 million and is owned by a Canadian media company. In its Lasipalatsi studio, it produces several TV programmes, among which is a very popular youth programme. The studios are at floor level and have large glass windows, so (live) recordings of popular shows can be watched by people passing by. Although the recordings of live shows take place in Lasipalatsi, the actual broadcasting facilities are located in the outskirts of Helsinki, some 5 km away in the Pasila area. The two locations are connected by high capacity data lines. The concept of open studios in inner cities was copied from the Canadian mother company of the production firm.

The firm choose the inner city location of Lasipalatsi to be close to the people and to stress the broadcaster's openness and liveliness. Another advantage of the inner city location is the nearness of media agencies and the media centre of the new Sanomat building, where the two biggest newspapers in Finland and many media firms are located. The live shows of MTV3 have greatly contributed to the image of the Laipalatsi building.

The broadcaster's activities are expanding rapidly: it started a new channel in February 2000, and became active in digital TV broadcasting (a local digital station for the Helsinki metropolitan area) in August 2000.

... in an Innovative Organisational Setting

It is not only the concept of Lasipalatsi that is innovative: the same holds for its organisational structure. Lasipalatsi is one of the first projects (together with the Cable factory and the Tennispalatsi) where the Helsinki City Administration has opted for an arm's length structure of governance, a new form of public/private partnership. The real estate department of the City Administration owns the land and the building (it owns 60 per cent of the land in the Helsinki area). The city government has outsourced the management and exploitation of the complex to an independent company named Lasipalatsi Media Centre Ltd, of which the city owns all the shares. The tasks of Lasipalatsi Media Centre Ltd are: 1) to lease the spaces to appropriate tenants who fit in the concept; 2) to collect sufficient revenues from rents and other sources; 3) to create synergy in the building, i.e. to coordinate the operations of the tenants, initialise and harmonise the projects; and 4) to see to the marketing of the Lasipalatsi. The board of this company consists of public and private tenants of the building and representatives of the city of Helsinki. An example of a technical innovation realised by the company is the home page machine, enabling every visitor to create a personal home page.[1] Another example is the book-on-demand service in the bookstore.[2]

In the decision-making process on the organisational form of Lasipalatsi, several solutions were considered:[3] one was to organise the project in the traditional way, by putting the management of the centre in the hands of existing departments of the city. Another option was to create a new administrative public structure to run the Lasipalatsi. Both alternatives were considered inappropriate. Keeping the exploitation and control of Lasipalatsi in existing city departments would have the disadvantage of complicated and slow decision-making and difficulties in sharing of responsibilities, because many departments are concerned; it would have become very difficult to create a well-defined image of the centre and the private partners would have to contact many different administrative parties. The second option, creating a single administrative city media centre office, would make for faster decisions and create some kind of profile for the centre. However, it would have been difficult to place the centre within the city's administrative systems and would have required an increase in city personnel; finally, the Media Centre's task is not administrative in nature, which makes an purely administrative solution less desirable.

Because the Lasipalatsi is a European Urban Pilot Project, yet another body was established, to guide the realisation of the project and to supervise and control the European Urban Pilot project. This was the Management Group, created for the duration of the European project, chaired by the director of the Helsinki City Library. The other members are from the City Office, the Real Estate Department, the Cultural Centre, Youth Affairs, Urban Facts, and Helsinki Festivals. In 2000, when the urban project came to an end, this group was dissolved.

The organisational form of Lasipalatsi has entailed a number of learning processes. First, the board members of Lasipalatsi Media Centre Ltd consider its mixed composition (public and private) an important learning experience: members representing the public sector have gained a better understanding of business needs and learned that 'things have a price', whereas the eyes of private members have been opened to the general interest. On a higher administrative level, the city has learned how to deal with public/private cooperation and gained experience on how to set up an arm's length construction.

... and Innovatively Financed

Lasipalatsi is one of the first projects in Helsinki in which European funding was involved (Finland became an EU member in 1995). In July 1997, the EC DG 16 announced that it had decided to assign EURO 2.7 million to the implementation of the project in the Urban Pilot Project Programme. This sum represents 30 per cent of the support-eligible total expenses of EURO 9 million. Two-thirds of this sum was used for renovation; the rest for special projects in the Lasipalatsi.

The Media Centre Ltd rents the building from the city of Helsinki for an annual EURO 1.2 million and collects the rents from the tenants. The rent levels are set in such a way that the Media Centre Ltd is able to fulfil its obligation to the Real Estate Department and has enough resources to start projects, maintain the infrastructure and undertake other activities in the general interest. The total budget of the Media Centre Ltd is EURO 1.7 million. Should total rent yield become higher than EURO 1.5 million, 50 per cent of the extra yield will have to be returned to the city.

Although the decision to renovate the building had already been taken before there was any question of EC support, European funding has been fundamental to the realisation of the concept. Without this support, the pressure to exploit it in a more profitable way would have been higher. One example of the contribution of the EC funding is the possibility of allowing lower rents for some companies. In the course of 2000, the Urban Pilot Programme came to an end. The challenge is to find new financial resources to substitute for the money from the Urban Pilot Project. The same holds to cover the high communication costs: the unexpectedly frequent use of the Internet facilities in Lasipalatsi during the first year costs EURO 134,000.

In Sum

It is widely agreed that the Lasipalatsi is a successful project: the number of visitors is much higher than expected, a number of innovative projects have been successfully executed and it clearly fulfils a role between the everyday world of citizen and the world of technology, by stressing the human side of technology. The organisational structure, new in Helsinki, has proved its value in several respects.

It is important to note that the City Council's decision to renovate the building was taken before European funding was granted: however, the money from Europe was decisive in realising the concept as it has evolved.

The EC project has not only provided financial support: it also 'obliged' the Media Centre Ltd and the city to create and maintain a concept. The end of the Urban Pilot will mark a new stage in the development of the concept: it means that financial support for special projects in the Lasipalatsi – which form a substantial part of the value-added of the concept – has come to an end. Solutions should be found to fill this gap. Increased sponsoring of projects could be an option.

Without a concept or long-term strategy, increasing market pressure might endanger the concept in the long run. The popularity of the building and the steeply rising rent levels in the city of Helsinki in general may make it more difficult to maintain relatively unprofitable activities in the building, even if they fit very well into the innovative concept. In other words, there will be tension between direct rent yields – which are the interest of the Real Estate Department – and the indirect benefits for the city centre of having an innovative media centre. To withstand the power of the market, it is becoming more essential for the management of Lasipalatsi Ltd to increase the coherence and the value-added of the concept for the liveliness of the city centre. The city should be clear on its intentions.

A survey has shown that many visitors have only a very limited idea of which functions are present in the Lasipalatsi building (Lasipalatsin Mediakeskus Ltd, 1999). Improvement of the 'internal accessibility' of the building, for instance by creating an internal routing system, could lengthen the visits and add to the value of the concept.

To develop and market the concept further, the board of Media Centre Ltd should be powerful and synergy-oriented: it should be able to take the lead in the concept, urge tenants to participate and cooperate, raise external funding and engage in external networks. Regarding this, it might be good to have a Board that does not only consist of tenants who have their own interest in the building (some of whom are not interested in synergies at all) but to include also specialists from outside who clearly consider the project in its entirety.

3   Lasipalatsi and the Spatial-economic Developments in Helsinki

Lasipalatsi is part of the very dynamic city centre of Helsinki. Where the last section focused on the project itself, here we consider its role in a wider spatial-economic context. What is the role of Lasipalatsi in the inner city and the wider metropolitan area? How does it relate to other projects in the region? How does the Lasipalatsi fit into the urban strategy of the city and the region?

Developments in the Helsinki Metropolitan Area

The Helsinki Metropolitan Area (consisting of Helsinki, Espoo, Vantaa and several other municipalities) is one of the fastest growing urban areas in Europe. The population (about 1.2 million) is increasing fast, as in the other large Finnish cities, mainly owing to the immigration of (young) people from small towns and the countryside.

The economy of Helsinki (and of Finland as a whole) is booming, after a severe recession during the early 1990s, when the important export market, Russia, collapsed and Western Europe went into recession as well. The unemployment rate in the Helsinki region reduced considerably between 1991 and 1999, from 16 per cent to 8 per cent. The economic boom of recent years has increased rent levels in the city centre by 10 per cent annually.

One important engine behind the economic recovery is the tremendous growth of the information sector, particularly in telecommunications, business services, including data-processing, legal activities, business and management consultancy and advertising. This sector provided two-thirds of the growth in jobs in the Metropolitan Area in the business sector (City of Helsinki Urban Facts, 1999, p. 16). Within the information sector, the media sector (the target sectors of Lasipalatsi) have grown substantially in Helsinki as well (see Table 7.1). The table also shows the dominance of Helsinki within Finland in the media sector.

The Place of Lasipalatsi in Media Value Chains

The boom in the media sector in Finland is evident: it is also clear that Helsinki benefits more than proportionally, given Helsinki's share of Finland's media

Table 7.1   Employment (growth) of the media sector 1993–97, share of Helsinki in Finland in 1997

	1993	1997	% change	Helsinki's share of Finland
Publishing	4,464	5,107	14.4	38.7
Market research/public opinion polling	331	566	71.0	61.9
Advertising	2,345	3,171	35.2	57.7
Motion picture and video activities	769	903	17.4	60.2
Radio and TV activities	3,754	4,121	9.8	70.1
News agencies	325	327	0.6	79.0
Total	11,988	14,195	18.4	

Source: City of Helsinki Urban Facts (1999).

content production. To determine the place of the Lasipalatsi project in this respect, it is useful to take the *media value chain* as tool to shed light on this issue. A value chain represents the several stages of production, from the initial development stage to the final consumption of a product or service. Although usually this concept is applied to industrial sector analysis, the production of media products and services can be represented in this way as well. In Figure 7.1, the media content value chain is schematically represented.

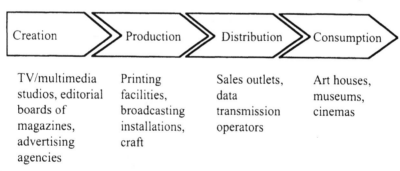

Creation	Production	Distribution	Consumption
TV/multimedia studios, editorial boards of magazines, advertising agencies	Printing facilities, broadcasting installations, craft	Sales outlets, data transmission operators	Art houses, museums, cinemas

Figure 7.1   The value chain of media content

All of the stages of the media content value chain are present in the Helsinki Area, but, unlike the past, each function has different locational preferences. The spatial separation of content development, production and distribution has been made possible by new data transmission possibilities.

In the content *creation* stage, the city centre plays an increasingly important role as location for the media industry. More and more, news makers and other content creators prefer the inner city, as this is an information-rich environment in which this sector can flourish. Another advantage is the nearness of the parliament and ministries, important 'news generators'. The move of the studios of public and private broadcasters to Lasipalatsi is illustrative of the attractiveness of the city centre. Recently, the largest Finnish newspaper *Helsingin Sanomat* opened its head office in the city centre, very close to Lasipalatsi (on the other side of the street). This building, named 'the new media house', is (partly) open to the public; it contains several bars and restaurants and hosts a number of other media firms as well. The *production* and *distribution* of the media content has a far weaker link with the city centre: the printing facilities of the newspaper Helsingin Sanomat for instance are located outside Helsinki, where land is much cheaper. In parallel, the actual broadcasting of MTV3 (one of the private TV stations of Finland, located in Lasipalatsi) takes place from Pasila, 5 km from the city centre. In the *consumption* of media content, the city centre is important again, as the location of cinemas and art houses.

Thus, in terms of the value chain, Lasipalatsi forms part of a 'media content cluster' in Helsinki's inner city. Also, it has an important role in consumption, since one of its principal goals is to bring new media and ICT to the ordinary people, in a very accessible way. Finally, Lasipalatsi has a function in linking consumption with creation (see also section 2): the presentation and testing of new technologies to the general public in Lasipalatsi fills the gap that often exists between the two ends of the value chain.

Links of Lasipalatsi with other Focal Centres of Media and ICT in Helsinki

This observation opens perspectives on the relation of Lasipalatsi with other media- and ICT-related centres in the Helsinki Metropolitan Area. Helsinki is a very strong centre of research and development in the field of ICT and new media, even in a global perspective, with Nokia as undisputed flagship. There are a few important locations in the region where the bulk of these research activities are located. The principal 'focal centres' are the so-called 'Art and Design City', and the information technology cluster in Espoo. Both centres are very strong in creating new technology and new ICT applications and services: they host firms and universities that operate at the forefront of technological developments. In the box below, more details are given about these centres. They were created with a lot of support of the local government, at a time when the Helsinki city had decided to become a technology and knowledge-intensive region.

New media development: Helsinki's Art and Design City
The Arabiaranta site in Helsinki is located at 6 km from the centre of Helsinki. This large site (about 85 hectares) is intended to become a so-called 'Art and Design City' within the city of Helsinki. A few years ago the City of Helsinki and the other owners of the Arabia district inaugurated a large development scheme called the Art and Design City Helsinki Project, aimed to turn Arabianranta into one of the leading art and design areas in Europe, where the functions of living, working, studying and recreation are mixed. It will become an important centre of experimental media production and testing of new media possibilities The project is planned to be completed by 2014. The area hosts already 4,000 residents and 3,000 jobs; these numbers are intended to increase to 12,000 jobs and 8,000 residents. The University of Art and Design is also located on the site, and offers research facilities, attracts many young people and acts as a breeding ground for young talents and new activities. In early 2000, an audiovisual complex was completed near the university (EURO 29 million). As an integral part of ADC new high-quality housing and shopping facilities are being constructed. The electronic infrastructure of the Art and Design City will be very sophisticated, with a region-wide broadband fibre network and open access for all residents, schools and businesses.

ICT development: Otaniemi in Espoo
The major site for ICT development is Otaniemi Technology Park, located in Espoo, a municipality west of Helsinki (but part of the Helsinki Metropolitan Area) at some 20 minutes drive from the centre of Helsinki and the Helsinki international airport. In the science park, about 5,000 researchers are employed, working for institutes of advanced and applied research or for one of the nearly 200 companies in Otaniemi Science Park. The main building of the Park is Innopoli, which is one of Europe's most extensive commercial science park centres. In Innopoli the interplay and cooperation of science, product development and business activities takes place. The aim of Innopoli is to speed up the commercialisation and internationalisation of the results of investments in Finnish research and product development, mainly in ICT. The site hosts over 100 companies. Annually, some 20–30 high-technology companies are created at the site.

For Lasipalatsi, the presence and proximity of these centres of competence could form an important advantage. Currently, there is no cooperation between the Lasipalatsi and these two concentrations of R&D in the field of media and ICT. Strategic cooperation between Lasipalatsi and these centres could benefit both parties: with its strategic position in the city centre and its public function, Lasipalatsi could function as showcase and test-bed of the new inventions from these areas, and could help the technology developers to become more customer-oriented – an increasingly critical issue in ICT and new media.

Lasipalatsi and Urban Strategies

To what extent does the Lasipalatsi project fit in a wider vision on and strategy for the development of Helsinki? The severe economic crisis of the early 1990s did not bring only misery. It was an important stimulus for the actors in the region to join forces and combat the crisis; momentum was created to achieve fruitful cooperation. As a result, in 1994, the key actors in the region – administration on different levels and the business community – developed an integral and broadly-shared vision and drew up a regional strategy. Key elements of this strategy were to develop Helsinki as a Nordic region of interaction, with a focus on sustainable, knowledge-based development, culture and information. Many plans were developed to give substance to this strategy. Examples are the setting up of science parks and the Art and Design City; the investment in Lasipalatsi as a media centre and centre of dissemination of new technology clearly fitted in this broad strategy as well, and contributed to its realisation. However, there were also less strategic reasons why the plan was realised: the high unemployment at that time provided an additional argument in favour of renovating the building to create jobs in the construction sector.[4]

Inner City Strategy

In the strategic policies for the development of Helsinki, the inner city receives special attention. In the mid-1990s, it was widely felt that the city centre should become a more lively area, for economic and cultural reasons. More specifically, a 'living room' function was felt to be missing: the city has a clearly demarcated shopping district and an administrative district, but lacked a 'social leisure district' where people could meet, eat, drink, consume culture, etc. This was felt as deficiency, particularly in a time of rapidly increasing cultural consumption (see Table 7.2).

Since then, much has been invested in developing a part of the inner city as the 'living room' of the Helsinki population, and making it a lively centre of cultural production and consumption, shopping and living. To create a mix of functions was considered fundamental from the beginning, to ensure liveliness all through the day.

Table 7.2   Growth percentages of domestic household expenditure on admission tickets between 1990 and 1995

Cultural consumption	Growth
Museums, art exhibitions	39.0%
Films, film clubs	15.0%
Theatre, opera, concerts	19.5%

Source: City of Helsinki Urban Facts (1999, p. 15).

Investments have been particularly directed towards this target. The best known example is the Kiasma, a progressive Contemporary Art Museum opposite Lasipalatsi. Another example is the reopening of Tennispalatsi, near Lasipalatsi. This building earns its name from the four tennis courts which were set up on the roof for the 1940 Olympics. Demolition has been on the agenda since 1957, when the building passed into the hand of the city. It has now been turned into a centre for cultural life: it was reopened in 1999 and hosts cinemas, a museum of cultures and additional premises for the Helsinki City Art Museum (City of Helsinki Urban Facts, 1999, p. 15). All these initiatives illustrate that the new function of the inner city as leisure/cultural consumption area is clearly emerging. There are plans for a new music hall in this area as well.

Internationalisation Strategy

Since 1994, internationalisation is one of the policy priorities of the Helsinki city council. The main goal of this was to develop Helsinki as a strong centre serving the Nordic countries, the Baltic states and Northwest Russia. In 1999, the strategy was updated (Helsinki City Office, 1999). Priorities are to promote internationalisation of business, foreign investment, and stimulate urban tourism. In the field of leisure and culture, the city aims to host internationally important events in the fields of sports and culture.

Lasipalatsi, being the first Urban Pilot Project, has added to the internationalisation strategy in the sense that it has generated much useful experience with European projects and enabled many contacts and cultural exchanges between the Helsinki and other European cities. Also, the annual international film festival at Lasipalatsi helps to put Helsinki on the international 'cultural' map. Regarding urban tourism, Lasipalatsi plays a role as an 'attraction' for visitors. One drawback is the lack of English-language signs in the building. Lasipalatsi has no specific role in the international city marketing efforts of the city, despite its special position as showcase of innovations in the information technology and new media.

In Sum

The function of Lasipalatsi as a centre of content creation, disseminator of new ICT and cultural consumption centre perfectly responds to spatial-economic developments in the media sector: on the supply side, creative producers increasingly prefer inner-city locations, and on the demand side, more and more people go out to sample culture, have fun, and be entertained. It thus fits perfectly in the urban strategy of revitalising the inner city. The fact that other content creators have also moved to the city centre increases the synergies, as face-to-face contacts are essential in the information-sensitive creative activities.

Lasipalatsi should be seen in the wider spatial-economic development of Helsinki's new media and ICT industries. Strategic links with other centres in the city – e.g. the A&D city and the Otaniemi Techno-Park – could strengthen the role of Lasipalatsi as real showcase of new developments, and could also offer the developers of new technology the possibility to put their products to the test. The current capacity of Lasipalatsi to demonstrate new technology is limited to the HTC (Helsinki Telephone Corporation) shop and the Nykyaika. Perhaps this should be extended. It would further strengthen Helsinki's already strong image as a forward city in the information society. The form in which such cooperations could take place is open to discussion.

4   Conclusions, Lessons for other Cities and European Urban Policy

It is generally recognised that there is still a large gap between the possibilities of (information) technology on the one hand and the use and realisation of these possibilities on the other. Only a fraction of what is feasible is actually used. The Lasipalatsi project shows the next step in ICT development: bringing ICT and everyday life into contact with each other.

In the fast-moving field of technology and media, Helsinki is one of the most advanced cities in Europe with regard to both new ICT development and adoption. The Lasipalatsi project is the innovative means to stay at the forefront, by bringing technology to the people (making it easy), better adapt it to the needs of the people (making it useful), in a leisure-like setting (making it fun). In this respect, it forms an important good practice project for other cities that also have the ambition to take a step forward in the information society.

The organisation of such a complex and multifaceted project as Lasipalatsi needs a public/private approach. The experience of Lasipalatsi shows that the arm's length organisational form is appropriate: it provides for the much-needed speed of action and enables new developments to be dealt with in a very flexible way. Particularly in this fast-moving and ever-changing field, freedom of action and flexibility are preconditions to make a project such as the Lasipalatsi a success. This implies that a different urban management style needs to be developed. Urban managers need to become more aware of technology development, but also of the needs of the businesses in general. They should be able to engage in networks with private partners and make use of their knowledge and resources, but at the same time keep the general overview of the public interest.

For the near future, the challenge will be to extend and further develop the concept: a strong vision on the building and its surroundings is needed to withstand short-term market pressures. The Real Estate Department, one of the principal decision-makers on the project, should withstand the seduction of increasing direct revenues but also consider indirect benefits of Lasipalatsi as inner-city attraction, and the potential of the centre to strengthen the local ICT sector and speed up the diffusion of new technology.

To prove the value of the concept, the Lasipalatsi Media Centre Ltd should continue to stimulate synergies, and therefore needs strong management to execute this task and develop projects. The showcase and test functions of the centre in particular could be enhanced by cooperation with centres of excellence in the region such as the A&D City and the Otaniemi Technology park.

The Lasipalatsi shows how European policy can contribute to the realisation of innovative projects, at relatively low costs. In Helsinki, the European urban pilot scheme was a trigger to redeveloping a historical building, experimenting with new governance structures and, particularly, creating the concept of a new media creation and diffusion centre. The bottom-up approach of the Urban Pilot project – cities had to come up with good projects and received funding if they were considered innovative enough– has worked out very well, but has been very limited in its resources. Still, the positive experience offers scope for a more integral European support for larger opportunity-based integral urban projects.

Notes

1   It works like a photo-machine: The machine takes a picture and lets the user fill in a few text files (personal data, hobbies) and use a style. Costs are EURO 3.3 only.
2   The objective of the project, in which Xerox Ltd was an important partner, was to test possibilities of new digital printing technology in publishing. The service has a selection of out-of-print titles and other books that are hard to get. The are printed, cut and bound on location, on demand, in 20 minutes time (the big Finnish publishing houses were not very cooperative in this project).
3   The city had established a working group to find an appropriate organisational structure for the project.
4   Looking back, in the early 1990s the conditions to restore the building were favourable: the severe recession in Finland and correspondingly low wages made construction relatively cheap. Furthermore, the high unemployment rate provided an additional argument in favour of reconstructing the building to create some employment.

References

City of Helsinki Urban Facts (1999), *The Information Sector as the Flywheel of Economy in Helsinki*, City of Helsinki Urban Facts.
City of Helsinki Urban Facts (1999), *Arts and Culture 1999*, City of Helsinki Urban Facts, statistics.
Culminatum Ltd (1998), *Background Material for the Urban Pilot Project Lasipalatsi Film and Media Centre*, Annual Teport.
Helsinki City Office (1999), 'From new openness to everyday normality: new policy priorities for Helsinki's international activities', memorandum.
Kanninen, V. and Mäkelä, L. (1999), 'Lasipalatsi – Research', draft, City of Helsinki Urban Facts.
Lasipalatsin Mediakeskus Ltd (1997), *Urban Pilot Project Lasipalatsi Film and Media Centre Helsinki*, Annual Report.
Lasipalatsin Mediakeskus Ltd (1999), *Annual Report: Urban Pilot Lasipalatsi 1999*.

Discussion Partners

Mrs M. Berndson, Helsinki City Library, Library Director.
Mr L. Helminen, Lasipalatsi Media Centre Ltd., Executive Producer.
Mr E. Holstila, City of Helsinki Urban Facts, Managing Director.
Mrs M. Kajantie, Lasipalatsi Media Centre Ltd, Chairman of the Board.
Mr V. Kanninen, private consultant.
Mr T. Karakorpi, Helsinki City Office, Director of Steering Committee INFOCITIES-Helsinki, IT Manager.
Mr T. Karttaavi, Helsinki Telephone Corporation Multimedia Division, Project Manager.
Mr E. Kauranen, Helsinki Telephone Corporation.
Mr T. Korhonen, Helsinki City Office, Financial Director.
Mr M. Kulmala, MTV3 New Channels Division, Director.
Mr K. Mikkelä, Nykyaika project, SITRA, Project Manager.
Mrs M. Raunila, The Cable Factory Cultural Centre, Director.

Chapter Eight

Manchester

1   Introduction

The city of Manchester is the capital of England's northwest region. It numbers 440,000 residents and forms the heart of the Greater Manchester area, with a population of 2,578,000. Manchester, the cradle of the Industrial Revolution, has had to cope with severe economic restructuring as the manufacturing sector started to decline from the late 1960s onwards. The city has experienced severe problems, in common with many other major industrial cities in Europe and the US. Between 1975 and 1990, Manchester lost over 100,000 manufacturing jobs and approximately one-third of its population. In the same period the service sector created many new jobs (90,000), but the problem was that most of these new jobs did not go to the people displaced by the job losses in manufacturing. For years, the economic performance of the greater Manchester area has been below the UK and European averages in terms of gross regional product and unemployment.

Like many other European cities, Manchester is a divided city in some respects: the problems are concentrated in some deprived areas, where employment rates exceed levels of 20 per cent and drug/alcohol abuse and crime abound. Manchester's growth areas are its revitalised city centre and the area around the airport to the south of the city. Both are at walking distance of two of Manchester's worst areas for poverty and unemployment (Carter, 1997). From this perspective two challenges stand out for Manchester: first, how to create conditions to achieve sustainable economic development in the region, to create more jobs and increase regional income, and second, how to tackle the social problems, concentrated in some neighbourhoods.

In this case study, we focus on the question how ICT offers new opportunities for both economic and social regeneration in Manchester, and how urban management could grasp them. In section 2 we take the economic perspective: in what way can Manchester benefit from the current growth of the ICT sector; what are Manchester's strengths and weaknesses? Also, we describe and evaluate regional policies to further the ICT cluster. Section 3 is about ICT and social inclusion. Here, we address the question of how ICT can contribute to the inclusion of less favoured groups into the information society, and show some examples of projects in Manchester in this field. In section 4, we draw some conclusions and present some policy recommendations.

2   The Manchester ICT Cluster

ICT is the principal growth sector of the early twenty-first century. Its development is driven by super-fast technological evolution, resulting in a continuous stream of new products and applications onto the market, eagerly used by companies and consumers. Another important growth impulse for the ICT sector is the liberalising of telecom markets in Europe, which entails new players, lower prices, more competition and a much accelerated innovation process.

Notwithstanding the capacity of ICT to demolish time and space barriers, big cities are leaders with respect to the development and application of ICT and the construction of infrastructure (Alles et al., 1994; Graham and Marvin, 1996). Important segments of the sector are relatively footloose, i.e. not bound to specific locations, leading to interurban competition for ICT firms. That does not mean that ICT firms are generally indifferent regarding their location: some of them attach considerable weight to locating near their customers; this holds, for instance, for many ICT consultants, or for service departments of telecommunications providers. Large parts of the sector are thus a function of the general local economy. This implies that large cities, with their concentrations of 'demanders', are more likely to be attractive to ICT activity (Castells, 1996). Although in virtually every sector the demand for ICT products and services is strong (and growing), the most important demanders are the financial sector (banks, insurance), and the media sector.

A major location factor is the availability of appropriately skilled staff. This gives cities with a high quality of life a head start:

> To win at this game in the long run, they [cities] will need the right sorts of local attractions to retain the talent –in particular, pleasant and stimulating local environments high-quality educational and medical services, and sufficiently flexible transportation infrastructures and building stocks to accommodate rapidly reconfiguring patterns of activity (Mitchell, 1999, p. 111).

Some firms have special needs with respect to the quality of the information infrastructure; this holds for companies that transmit enormous flows of data, such as Internet providers or electronic publishers or other media companies.

They prefer locations where this infrastructure is in place.

In sum, there is a wide variety in locational needs in the different segments of the ICT sector. How attractive is Manchester as a location for ICT firms? What are its chances to benefit from the sector's growth? What and how can policy makers contribute? This section sketches Manchester's position as cluster of ICT activity. Also, local and regional policies that aim to further strengthen the city's cluster are described.

Manchester Hosts a Well-developed ICT Business Sector

The city of Manchester is the principal centre of ICT activity in the northwest region of England (BT, 1999). The ICT functions undertaken in Manchester range from the manufacture of electronic components and networks systems hardware to provision of technical support and all kind of services; most firms predominantly operate for the regional market, but in some respects the cluster has a function that supersedes the regional dimension. For instance, the region hosts some leading ICT firms' European headquarters. The city also hosts around 1,000 smaller ICT firms active in software development, Internet and telecom firms; their number has been rising sharply in recent years, as is reflected in the development of the Manchester Science Park.[1] In the northwest region, 30,000 people are employed in the ICT industries (MIDAS, 2000), a substantial number of them in Manchester.

Types of ICT business in Manchester
European headquarters: ICL/Fujitsu (1,500 staff), Siemens (Energy and Automation Division) 500 staff, Brother International Europe (350 staff), Sharp (UK) 400 staff.
Main regional operation: IBM, Cap Gemini, Sun, Bull, Logica, ICL, Oracle, Sema, Andersen Consulting, HP.
Fast-growing software companies: Infogames, CMG, Admiral, S1.
Fast-growing Manchester Internet companies: Powernet (Internet Infrastructure), TeleCity (Internet services), Kewill (E-commerce), XTML (Internet infrastructure), Harbinger (e-commerce solutions), Burns Open Systems (e-commerce).

Spatially, the ICT sector is strongly concentrated in the centre of the city (the high-quality central location for business), the Manchester Science Park (near the university and city centre), and in the southern part of Manchester. This last location is not only popular because of the proximity of the airport but also due to the high quality of housing in this area; it is one of the wealthier quarters of the town, with lots of green zones. The science park is also becoming more and more attractive for ICT firms. Since 1998, TeleCity, the UK's only Internet exchange outside of London, has been located there. This puts the area at the heart of UK Internet infrastructure development by offering the second major UK hub to the transcontinental Internet backbone (the other network access point in the UK is located in London). Businesses wanting to lease high speed Internet connectivity can now take advantage of cheaper rates, as access is direct rather than via London. This complements the Manchester Network Exchange Point (MaNAP) set up by the University of Manchester in 1997 and together these services have been a key factor in the rapid growth of the Internet and e-business sector in Manchester.

What else explains the favourable development of (international) ICT business in Manchester? A recent Manchester ICT survey[2] indicates a few strong points: close proximity of ICT suppliers, excellent transport links, access to markets via Manchester International Airport, the availability of skilled labour, the availability of office space, and the low level of wage rates. The international airport, located at the southern edge of the city, is of utmost importance for Manchester's ICT cluster, as not only the large ICT firms but also many smaller ones are internationally active.

Manchester's strength as location for ICT firms can also be related to its economic structure: see Table 8.1. Particularly the large financial sector and other services generate much demand for ICT products and services. Furthermore, the local media sector is a driver behind the sector, both as demander of ICT services and as source for new media/ICT combinations. The broadcasting media sector is deeply rooted in the greater Manchester area: the BBC keeps an office in the city, but more important is the Granada Media Group, part of the independent television network ITV. The region also hosts a number of smaller TV producers.

The Role of the Research and Education Infrastructure

A major strong point of Manchester is its universities. The region contains four universities, with a total student population of 77,316. In 1997, the four universities collectively produced some 1,000 graduates in computer science and IT; another 2,300 graduated in engineering and technology-related disciplines (see Table 8.2).

The benefits of these institutes of the region's ICT sector are manyfold: they 'feed' the regional sector with a large number of highly skilled staff; in

Table 8.1   Employment in the Manchester TEC area, 1995

Sector	Employment	%
Agriculture and fishing	230	0.0
Energy and water	4,650	0.9
Manufacturing	81,030	16.0
Construction	18,860	3.7
Distribution, hotels and restaurants	102,170	20.2
Transport and communications	41,530	8.2
Banking, finance and insurance, etc.	104,140	20.5
Public administration, education and health	136,480	26.9
Other services	17,860	3.5
Total	506,950	100.0

Source: AES (1995), in Manchester TEC (1998).

addition, more and more students create their own companies after graduation, contributing to the important indigenous development of the region. Also, staff from the universities seem increasingly inclined to start new businesses. More indirectly, the student population greatly contributes to the liveliness of the city, thereby increasing its attractiveness for business activity.

Table 8.2   Manchester's universities, student numbers, and IT graduates

Institution	Total number of students, 1997/98	Computer science and IT degrees in 1996/97
Manchester Metropolitan University	28,362	169
Victoria University Manchester	23,571	317
Salford University	18,603	185
UMIST	6,780	328
Total	77,316	999

Source: Higher Education Statistical Agency (1998).

Despite this, we found that the contribution of the universities to the urban economy could be much greater, in two respects: first, university–industry relations could be improved, and second, the number of spin-offs from the universities could be much greater. To achieve both, universities need to adopt a more entrepreneurial attitude. The two issues need some elaboration.

University–Industry Relations

Potentially, the university could have great value for local industry, which could benefit from the knowledge embedded in the universities. However, incentives often work in the opposite way: for universities, what counts most is the volume and level of scientific publications, as these indicators determine the national and international ranking and status. Applied research for the industry is valued much lower. This implies that knowledge and technology transfer from universities to local business is low; the knowledge 'embedded' in the institutes is hardly transferred to the local industry. A striking case is the transformation of the polytechnic into a university (the Manchester Metropolitan University) in the 1980s. The former polytechnic used to have many links with business and execute a lot of applied research, to the benefit of both the polytechnic and its industry partners. In the last decade, there has been increased pressure on universities to shift from applied to more academic research.

Start-ups

In the ICT sector, business start-ups play an important role; many of the now-large ICT companies such as Amazon.com were started from scratch by entrepreneurial youngsters in the 1990s. In the USA, a substantial number of successful ICT firms started as spin-offs from a university and grew tremendously in only a few years. The number of start-ups from the universities in Manchester (as in all Europe) is low compared to the USA, for many reasons; the entrepreneurial attitude of students and university staff is far below American levels, although the situation is improving. Risk-taking is much less common. An institutional barrier is that patents are owned by the university, not by individual persons, providing no incentive for commercialisation; finally, academic status is not derived from successful business but from scientific publications.

A critical project for exploiting the resources of Manchester and its universities is the Campus Ventures (CV) initiative. CV is a university-based firm – but financially independent form the university – that aims to increase the number of start-ups by the universities (see box below). This initiative helps start-ups in a structured and integrative way, provides incentives for commercialisation of good ideas, and ultimately contributes to the creation of an entrepreneurial culture at the universities.

Good-practise incubation initiative: Campus Ventures
Campus Ventures (CV) offers space, legal support and financial support, and provides access to university facilities such as labs or powerful computers. CV is funded with ERDF and ESF funds, by national schemes and royalties: start-ups have to pay royalties to Campus Ventures as a certain percentage of their turnover, after three years of profitable operation. In the last five years, 50 university start-ups were supported by CV, offering 200 jobs (*Campus Ventures News*, 1998; interview). Some of them are expanding rapidly. Total 'public' funding of CV amounts to an annual EURO 2 million.

CV started at the University of Manchester, but has now spread its wings to the other universities in the region, through a firm named 'incubation partnerships'. CV has also links with the Science Park near the university; there are common board members. This facilitates the move of expanding firms from the university to the science park.

A related initiative to stimulate entrepreneurship from universities is the idea to start a 'Science Enterprise Centre', an education institute that should offer postgraduate courses on entrepreneurship.

Economic Development Initiatives

Manchester's ICT cluster is developing favourably thanks to the generic growth of the sector and Manchester's attractiveness for that type of activity and its rich knowledge base. Nevertheless, public administrations (on many levels) take much effort to stimulate the local and regional ICT cluster even further. The most active organisations in this respect are the City Council, MIDAS (Manchester Investment and Development Agency Service) and Manchester Enterprises (a joint venture between the city council and the chamber of commerce). The following initiatives have been taken:

- *Attracting ICT companies from abroad.* MIDAS (the inward investment agency for the City Pride area[3]) has designated ICT as a spearhead sector. It aims to attract ICT and Internet companies to the region mainly from the USA and Canada, by offering them information on the area and all kinds of support (appropriate location and skilled staff). Also, MIDAS supports Manchester-based rapidly expanding firms in finding appropriate locations and helps firms to obtain available subsidies and grants. The organisation is funded by four municipalities and ERDF funds.[4]

Interestingly, although it covers four municipalities, MIDAS uses 'Manchester' as brand name to sell the whole City Pride area.

* *Helping SMEs with the adoption of ICT.* Many firms find it difficult (or do not see the need) to apply ICTs in their business processes and are therefore threatened by loss of competitiveness. The City Council therefore supports several initiatives to help SMEs in their transition to the information economy. An example is its support to the Manchester Institute for Telematics and Employment Research (MITER), a university-based organisation that helps SMEs to apply Internet technologies and e-commerce, or link them to the appropriate business partners. Thirty-five per cent of MITER's budget is ERDF-funded, channelled through the City Council. Under the umbrella of MITER, another project in this field is the ISaware (Information Society Awareness Project) initiative,[5] aimed at helping SMEs adopt ICTs and e-commerce. Priority is given to SMEs in regeneration areas in the city, and the adoption of ICT by firms in the tourism sector (Manchester City Council, 2000). A similar initiative, aimed to support SMEs and start-ups, is Business Link. This programme runs in all English regions. In Manchester, Manchester Enterprises is responsible for the programme.

* *Stimulating start-ups in ICT.* The incubation activities of Campus Ventures (see above) are supported by the City Council, as Manchester's urban management realises the importance of incubation for long-term economic development.

* *Creating/supporting business networks.* Young ICT and new media companies are hardly represented in the traditional networks such as the Chamber of Commerce; they have different preferences and wishes regarding network formation and prefer a more informal setting. Therefore, the Manchester City Council supports an initiative to build a North West New Media Network, to bring new media/ICT firms together, let them share experiences and develop joint projects. MITER is responsible for the execution of the programme.

3   ICT and Social Inclusion

Social Problems

Like a number of other large European cities, Manchester has to cope with many social problems. In 1998, the city ranked third in the Index of Local

Conditions' list of most deprived districts in England (http://www.manchester.gov.uk/yourcc/final.htm#unemployment). To a large extent, these problems can be traced back to the restructuring of the urban economy. During the second half of the twentieth century many people lost their jobs because of the decline of the manufacturing industry. In the 1980s and early 1990s, national government cut the budgets of state welfare programmes, as well as that of urban regeneration schemes. This has left Manchester with a legacy of large areas of high unemployment, alcohol and drug abuse, crime, and a very poor housing quality. Many thousands of people live in relative poverty and are excluded from social and economic life. More recently, national government has adopted a variety of approaches to tackle problems of deprived neighbourhoods, from central–local government partnerships (Urban Programme) to top-down market-led solutions (Urban Development Corporations, Enterprise Zones), to the competitive 'challenge funding' (City Challenge, Single Regeneration Budget) approach (Mawson and Hall, 2000).

The New Labour government has not only committed more financial means to fight social exclusion; it has also introduced a policy known as 'joined-up government', focusing on an integrated approach to regeneration, with a larger role for local authorities as leaders of local communities. Local authorities will have the task of organising and supporting partnerships and guaranteeing quality services for all (IPPR, 1998). To fight problems of social exclusion, the Manchester City Council – often in partnership with other organisations – has initiated a wide array of policies, some directed to specific areas. Many projects and programmes are co-funded by European ERDF and ESF schemes.

The City Council regards ICT policy as one instrument (among many others) with which to fight social exclusion, unemployment and under-education. The basic aims of policies are to help disadvantaged people make the shift towards the information society, by facilitating access to ICT-related education and training, as well as access to the Internet. Also, regional and national government have launched all kinds of schemes and programmes to reduce the gap between the 'haves' and 'have nots' of the information society. The next subsection will describe some of the policy initiatives directed towards a wide array of ICT-related issues.

ICT-related Social Projects

Manchester counts a number of ICT programmes and projects for disadvantaged groups.

- *ICT education for specific groups.* The Electronic Village Halls (EVH) initiative offers access to Internet and information technology education and training for weaker social groups. The EVHs (started in 1992) offer Internet terminals, where the Internet can be accessed free; also, numerous training and education programmes are offered at low fees. The EVH initiative is supported by the European Social Funds and ERDF, as well as the City Council. Manchester counts three EVHs: one of them is area-based, and functions in the borough of East Manchester; the other two are directed towards specific groups, namely minorities ('the Bangladesh house') and women (Women's Electronic Village Hall): see box below.
- *ICT education in local communities.* There are several local community and volunteer initiatives in the field of ICT. Some churches are active in ICT initiatives; for instance in the deprived area of East Manchester, a minister has set up an ICT-education scheme for children who have left school, as well as a centre for ICT education. ICL (a computer manufacturer) has provided the equipment.

Case study: The Women's Electronic Village Hall (EVH)
The Women's EVH is located in central Manchester, and is a 'women-only' centre for education in ICT. Weekly, some 100 women attend a variety of courses, ranging from basic computer skills to programming, Internet applications and web design. It was felt useful to set up a dedicated facility for women. Training methods are designed in such a way that they are attractive for women, with small classes, flexible working hours and support in finding child care. The centre attracts women from all over Manchester, with a variety of backgrounds. This diversity is judged very positively by the participants and results in a mix of social groups. Most women see the courses as an opportunity to increase their chances of getting a job. The Women's Electronic Village Hall has a total annual budget of EURO 561,000, provided by the Further Education Council, local colleges (the Manchester City College) and by European programmes. There is no private money involved, but sponsors are sought to enable a move to a new location in the Northern Quarter of Manchester.
Source: Women's Electronic Village Hall (1999).

- *IT equipment provision.* One example is ITEM (Information Technology East Manchester), a computer-recycling project initiated by the City of Manchester. The aim of this project is to create both temporary and permanent employment and training through the innovative treatment of waste computers and to provide IT equipment and services to the community. The organisation collects old computers, decomposes them, rebuilds new computers with useful components, and sells waste to a recycling company. In this way, computer equipment is provided to organisations that would normally not be able to obtain it. The project is a partnership of community groups, Manchester TEC, and regeneration agencies. It is predominantly publicly financed with Manchester TEC as the principal funder of the project; ERDF also contributes heavily (Manchester City Council, 1999).

• *Free Internet access points.* Many people do not have a PC at home, and thus have reduced chances to become acquainted with the new possibilities the Internet offers. As in many other cities, several public Internet access points exist in Manchester. The EVHs play a role here, but the libraries in Manchester in particular are active in facilitating access to Internet for everyone; each of the city libraries has a number of computers which can be accessed by the general public. To prevent people from downloading unsuitable material, each user has to sign a contract in which he/she agrees not to download this.

The fact that most initiatives were taken bottom-up can be judged positively, but a little more coordination between the initiatives could lead to better mix of training; other advantages could be resource sharing and good practice exchange. The establishment of links with more professional education/training institutes and links with private sector partners is also important. Therefore, the Manchester City Council has started to identify local-based initiatives and is actively seeking to bring more coherence in all the different approaches.

4   Conclusions

Manchester in the Information Society

Each city should find its own way into the information society, playing to its own strengths and opportunities. Manchester's ambitions are to develop a strong ICT business cluster, to make most out of its knowledge base and, at the same time, to enable weaker social groups to participate in the information society.

The city has the ingredients to benefit from the shift to an information economy. In the realisation of the first ambition – developing a strong local ICT business cluster – Manchester benefits from the explosive growth of the ICT sector in general. The city can boast a strong and diversified ICT cluster, supported by the rich local knowledge base. Much of Manchester's ICT activity is regionally demand-driven, as Manchester is the principal financial and service centre of the northwest region. A major strength of Manchester is the enormous creative potential of the city; its concentration of vibrant cultural industries in the production of content; the new media, particularly the Internet, opens new ways to offer and distribute creative products, even to a worldwide audience.

Many citizens are socially and economically excluded. There is a sharp contrast between the booming ICT cluster and the social and economic problems of large groups of Manchester's citizens who live in deprived areas; not only it is socially undesirable to exclude so many people from social and economic life, also from an economic development point of view this backlog is undesirable: labour shortages are beginning to emerge in the ICT sector, and deprivation-related crime poses a threat to the liveability of the city, reducing its attractiveness for firms. The cities' efforts to help people of weaker groups with ICT adoption should be considered appropriate, both from social and economic points of view.

Policy Evaluation

As shown in this case study, many ICT-related initiatives are running in the city. Many are initiated, supported or funded by public sector agencies, but developed and executed in partnerships, with other public administrations or with private or semiprivate partners.

Public/private cooperations abound. In general, the City Council efficiently makes use of existing knowledge, expertise and (financial) resources in the region by engaging in different kinds of public/private partnerships. One of the reasons is institutional: English cities have to compete for national resources and are obliged to engage the private sector in their strategies.[6] In many of the described projects, the Manchester City Council plays an enabling or supporting role, but 'outsources' the actual execution of programmes to existing organisations in the city. Examples are the ICT adoption schemes for SMEs, co-funded by the city council but executed by MITER (Manchester Metropolitan University), which has the relevant knowledge, or the many locally-initiated ICT initiatives such as the Electronic Village Halls. Positively, some officers of the City Council stay up-to-date with developments in ICT by intensely networking with local ICT firms, which leads to improved knowledge of the needs of the market, and, ultimately, in more effective policies.

More intensive private involvement could improve the formulation and execution of the region's inward investment policy: currently, the inward investment agency hardly involves private actors in its acquisition strategies. Also, in some community-based education programmes there is a lack of systematic private involvement, despite the fact that there are clearly private interests in supporting ICT adoption and training (such as a larger market for ICT services; potentially new staff; better quality of life and image of the city).

Public/public partnerships are badly needed. ICT-policy is a 'hot issue' in the UK. Many public organisations are in one way or another active in it: the Manchester City council, other city councils in the region, the Region North West, the national government and the European Commission. The question could be raised whether or not too many ICT-related initiatives are carried out, particularly in stimulating ICT adoption by weak social groups, and in helping SMEs with information technology.

Within greater Manchester, inter-municipal cooperation is smooth. The projects studied show that partnerships between public (or semi-public) actors in the Manchester TEC area is strongly developed. Most striking is the cooperation between the municipalities of in the inward investment, in MIDAS, that can serve as model for other European cities.

Most projects in our analysis were predominantly publicly financed. A larger share of private capital in some projects could be beneficial. A positive example is the cooperation of the SEMCCA (South East Manchester Church Community Care Association) with computer equipment providers to establish local ICT training possibilities.

Related to the above-mentioned issues, an important difficulty is how to optimally channel funding from all the different sources (local, regional, national and European). Most funds have their own requirements, making this an even more difficult task. Manchester being a European Objective 2 region, in most of the policy projects studied ERDF (European Regional Development Fund) funding plays a decisive role. One the positive side, this shows how important European funding can be to finance innovative policy projects. On the negative side, the question of what will happen when ERDF funding is cut can be posed. For existing and new projects, long-term sustainable funding solutions should be striven for to prevent their 'sudden death' after subsidies stop.

From these conclusions, a number of policy recommendations can be derived. First, the City Council could take the lead in guiding Manchester's way into the information society by leading the design of an ICT strategy for the metropolitan region. In this strategy, an integrative approach is needed, in which economic, social, education and mobility issues are all incorporated and seen in they mutual dependence. In this process, all the current initiatives should be involved, public and private partners consulted and involved in the development and execution.

A first step could be to make a listing of all the projects that run in the region, and evaluate them. Synergies among initiatives could be better exploited; the projects on the European national, regional and urban level should be harmonised. A more thorough listing of all the projects is needed to avoid duplication and wasting money.

The use of private resources (not only money but also networks, equipment and knowledge) for social initiatives should be encouraged; the benefits of the private sector are an increase in the educated work force, a larger market for ICT products and service and a better image. More private sector involvement is also desirable because of the probable future reduction of European funding of projects in the region.

The City Council should continue to support universities' initiatives, one of the city's major strengths in their incubation activity: the birth and growth of new firms, particularly in ICT, has a high payoff for the local economy. One possibility is the provision – under the conditions of safeguarding the value of the funds – of venture capital using pension funds. A second possibility is support for the setting-up of a school of enterprise, incorporating this into Manchester's marketing campaigns.

The efficiency and effectiveness of the inward investment agency could be improved by engaging private firms in its activities. Efforts could be targeted towards attracting specific firms instead of general acquisition; also, existing contacts of firms already present in the region could be exploited in this way. Lastly, the current focus on ICT firms could be supplemented by more focus on firms making heavy use of ICT, as an indirect demand stimulation for the sector.

Notes

1   Started in 1984 with one building, the Park currently consists of six buildings accommodating 45 technology-based companies, with around 550 employees.

Construction has commenced on a second site on the City-International Airport parkway, which can potentially double its size.

2   Midas (1999); 77 ICT companies participated.

3   This area covers the cities of Manchester, Salford, Trafford and Tameside.

4   MIDAS's total annual budget amounts to EURO 1,610,000, 50 per cent of which is funded by the municipalities. The ERDF fund contributes EURO 642,000.

5   This is a national programme. The project was initiated by the Economic Initiatives Group of the Manchester City Council, and works in the City Pride Area (Manchester, Salford, Trafford, Tameside). It is funded through ERDF.

6   Many of the initiatives described in this case study are co-funded through the Single Regeneration Budget, a national scheme integrating 20 different grants from five government departments. A substantial proportion of that is made available annually on a competitive basis to partners in the regions. For urban regions, the degree to which they manage to attract private capital is an important success factor in the competition for grants (Parkinson, 1998).

References

Alles, P., Esparza, A. and Lucas, S. (1994), 'Telecommunications and the Large City–Small City Divide: Evidence from Indiana cities', *Professional Geographer*, No. 46, pp. 307–16.

British Telecom (1999), *Telecommunications and Regional Development, Focus on the North West*, BT and the Local Futures Group.

Campus Ventures News (1998), Issue 4, Statistics, p. 3.

Carter, D. (1997), *Creative Cities and the Information Society*, Manchester City Council.

Castells, M. (1996), *The Rise of the Network Society*, Blackwell Publishers, Cambridge.

Graham, S. and Marvin, S. (1996), *Telecommunications and the City: Electronic spaces, urban places*, Routledge, New York.

IPPR Institute of Public Policy Research (1998), *Leading the Way: A new vision for local government*, IPPR, London.

Manchester City Council (1999), 'Information Technology East Manchester Refurbishment Pilot Project', project proposal.

Manchester City Council (2000), 'Greater Manchester Information Society Awareness Project', project plan 2000.

Manchester TEC (1998), *Economic Assessment 1998 – Part 3: Business and the Economy* and *part 4: Sectors*.

Mawson, J. and Hall, S. (2000), 'Joining it up Locally? Area Regeneration and Holistic Hovernment in England', *Regional Studies*, Cambridge, vol. 34, no. 1.

MIDAS (2000), *Environment for ICT*, internal publication.

Mitchell, W.J. (1999), *E-Topia: Urban Life, Jim – But Not As We Know It*, MIT Press, Boston.

Parkinson, M. (1998), 'The United Kingdom', in Berg, L. van den, Braun, E. and Meer, J. van der (eds), *National Urban Policies in the European Union: Responses to urban issues in the fifteen member states*, Ashgate, Aldershot.

Women's Electronic Village Hall (1999), Annual Report 1998.

Women's Electronic Village Hall (1999), Courses 1999/2000.

Discussion Partners

Mr D. Auckland, Campus Ventures, Managing Director.
Mr Dave Carter, City of Manchester, Economic Initiatives Group.
Ms C. Herman, Women's Electronic Village Hall, Project Manager.
Mr A. Jackson, Central Library, Project Manager.
Mr B. O'Neill, East Manchester Community ICT Network.
Mr T. Shaw, ITEM (Information Technology East Manchester), Project Manager.
Mr A. Young, Manchester Inward Investment Agency.
Mr T. Speake, Manchester Technology Management Centre, project manager, MITER Director.
Ms J.W. Seddon, North West New Media Network, Project Manager.
Ms Jane Bowdenleigh, Isaware, City of Manchester.

Chapter Nine

Marseilles

1   Introduction

The electronic revolution opens up great opportunities for cities: it offers solutions to reduce inefficiencies and bureaucracy within the municipal organisation and save money; it may greatly improve service delivery for the citizens; it can help to fight congestion and use infrastructure capacity more efficiently by providing better and faster information on the traffic situation; and last but not least, by improving the quality of the relation with citizens and solving all kind of problems, urban management can make a big difference to the attractiveness of their city as a place to live and work.

At the same time, implementing ICTs raises many new and sometime difficult questions. To list but a few: how to shift the municipal organisation to a demand-led organisation; how to reorganise and regroup information flows from many different departments; how to convince partners to join a certain information technology standard; how to reach all the citizens by using new media, and not only those with Internet at home; how to deal with new divisions of responsibilities between private and public actors?

This case study shows how the French city of Marseilles uses new technologies to improve service delivery to its population. Some examples of ICT projects are presented, which show how ICT can contribute to the solution of problems and the seizing of new opportunities, but also what technical and organisational difficulties may arise in these renewal processes. We try to draw some general conclusions, to learn lessons which can help other cities and to provide some recommendations for the City of Marseilles.

This case study is organised as follows: section 2 contains a brief introduction into the city of Marseilles. Section 3 describes ICT projects that are carried out in the Marseilles metropolitan area from both a technological and organisational point of view. Section 4 draws conclusions, derives lessons for other cities and provides some policy recommendations for Marseilles.

2   Marseilles: A Brief Introduction

Population and Economy

Marseilles, capital of Bouches-du-Rhône *departement*, is located in southern France on the Gulf of Lion (an arm of the Mediterranean Sea). With 800,550 (1990) inhabitants in the city proper, it is the second most populous city in France. Its metropolitan area population is 1,262,273 (1990).

It is a major seaport and an important commercial and industrial centre. The city is linked by canal with the River Rhône and is served by extensive rail and air transport; the large petroleum port of Fos, chiefly developed in the 1970s, is nearby. Products of the Marseilles metropolitan area include iron and steel, chemicals, plastic and metal products, ships, refined petroleum, construction materials, soap and processed food (Encarta.com).

Since July 2000, Marseilles, together with 18 cities of the Bouches-du-Rhône *departement*, has formed the urban community called the 'Métropole Marseille Provence'. This new structure has competence in the field of land management, housing, urban policies, service provision and environment protection. The intense functional and spatial relations between the city and the surrounding communities – commuter flows, traffic – now find expression in a formal structure, which will presumably facilitate the tackling of problems in the metropolitan area as a whole.

The economic situation in Marseilles is still worrisome. In the last quarter of the twentieth century, economic restructuring hit the urban economy, particularly the traditional manufacturing sector. Also, the loss of French colonies was a blow for Marseilles's trading function. Although many new jobs have been created in the tertiary sector, the city is still plagued by high unemployment rates, and incomes are low compared to other large French cities. The problems are aggravated by the influx of many low-skilled immigrants.

In the second half of the 1990s, several 'grand projects' were developed to improve the urban economic base and to increase the city's vitality. One of the principal projects is 'Euroméditerranée'. Within this urban renewal scheme, investments are made in infrastructure, real estate and neighbourhood renewal. It should help to differentiate the economy (attract new growth sectors such as information technology companies and service firms) and increase the number of corporate headquarters and other decision-making centres. Public investment in Euroméditerranée from 2000–06 amounts to EURO 237 million, provided by the state, the City and surrounding communities (Ville de Marseilles, 2000).

Marseilles's ICT Strategy

Marseilles ambitiously promotes itself a 'capital of new information and communication technologies'. The city has drawn up an ICT strategy, containing four dimensions: stimulation of economic development and territorial upgrading; improvement of public service delivery; the increase of Internet adoption; and the increasing use of ICTs within the municipality (Damlamian, 2000).

Encourage economic development The key objective is to facilitate and speed up the liberalisation of the local telecommunications market. This is expected to result in better quality of infrastructure, improved services and lower prices, which will in turn make the city more attractive as a location for business. The city has set out to facilitate the construction of new communications networks, to encourage new telecom entrants in the Marseilles market and to limit public investment. The city has designed a basic sketch of the electronic infrastructure that is needed, and lets the operators compete to construct and exploit. A one-stop shop agency has been created to welcome and receive new entrants. The city already hosts some 20 operators (Ville de Marseille, 2000).

Improve public services For companies, ICT greatly facilitates a shift from a product- to a customer-oriented approach. For local government agencies, a similar transition becomes possible. For an organisation based on competence domains, the citizen and his or her needs should be central. The city of Marseille aims to simplify the relation between city and citizen through an increased responsiveness of the public services and an increased differentiation, i.e. provision of different types of information to companies, tourists and citizens. To achieve this, in 1998 the city launched a new medium called 'teleservices à la population', an online service for the population. It offers economic, administrative and cultural/tourist information, an interactive service permitting the posing of questions or ordering of products. The services are accesible by PC and Minitel, but also at a number of public places.

Increase Internet adoption This is seen as essential for the city to move into the information age. Among other things, the municipality is investing in a three-year programme to connect 180 schools to the Internet and is creating access points in public places such as libraries.

Adopt ICT within the municipal organisation The ultimate end is to respond better to the demands of the public, to improve the quality of services and to reduce costs. The city is making the interconnection of public buildings a reality. Already, 13 sites are linked by a new dedicated infrastructure that runs through the metro tunnels. To create (and use) this network is cheap, as no digging is required. By the end of 2000 the network connected 53 sites. A further objective is to convert municipal information systems into a client-server system. An intranet is being developed. This should lead to better utilization of information resources and higher safety and protection levels.

3   Examples of ICT Initiatives

The ambitions of the municipality of Marseilles are high, but what has actually been achieved? This section lists some concrete examples of the deployment of new ICTs in the city and sketches organisational and strategic issues that surround the ICT projects. This section is predominantly descriptive; the lessons of the projects are drawn and generalised in the conclusions.

Transport: The Marseilles Smart Card

Like many other cities, the city of Marseilles suffers from severe congestion problems. They are caused by commuting flows, as many people work in Marseilles but live elsewhere. During rush hours, road capacity is insufficient to handle the massive car traffic from the suburbs and surrounding areas into the city. Another reason for congestion is the location of large commercial and shopping zones outside the city. Nevertheless, most people are reluctant to use public transport, one of the reasons being the lack of connectivity between the different transport systems (bus, metro, train) in terms of timetables and ticketing systems.

During the last decade, several initiatives were taken to improve the quality and interconnectivity of public transport in the metropolitan region, such as the construction of transfer centres at the edges of the city, where car users could park their cars and change to public transport. Also, a card was introduced that enabled payment for both the parking at the transfer points and the different types of public transport. This was already a great achievement, as many players had to be convinced to participate. At the track from Aix-en-Provence to Marseilles alone some 18 bus companies had to be brought together to cooperate.

Since 1996, a magnetic card system has been in use. There are three kinds of cards: a card for an incidental journey; a value card, with a value of FF100 or FF50, with which all kinds of journeys can be paid for (bus, train, metro, and parking payments); and personal cards, for people who travel at reduced rates, such as students and elderly people. Unlike the other cards, these are personalised. A central information system registers the sales of cards at the (many) selling points and the use of the cards in the various transport companies, and redistributes the revenues accordingly among the participating firms. At the time of writing, some 370,000 cards are in use. Compared to the previous situation (every company operating its own system) the magnetic card offers many advantages both to travellers and to the transportation suppliers. There are also drawbacks, such as frequent reading errors (due to damaged cards or reading equipment), and restricted information storage capacity (max. 128 bits/inch). Although is has not occurred yet, the card can be copied and is thus sensitive to fraud.

To overcome these drawbacks, the city transport department decided to develop a new card with a microprocessor, which would enable more information storage, allow more personalised and customised functions and be less fraud-sensitive. This also reduces reading errors: the card has a small transmitter that makes physical contact between card and reading equipment unnecessary. The card should contain four types of information: identification (personal data); the type of contract (reduced fee, fixed track, etc.); an electronic purse (the credit on the card); and memory space to track the card owner's 'travel history'. On the basis of this information, the automatic card readers at the transport nodes can determine how much to charge and deduct the amount from the card.

Organisational Issues

The introduction of a new system is never easy, especially when many players are involved. With the introduction of the magnetic card in the 1990s, the municipal transport department RTM managed to involve virtually all relevant actors, which was decisive in its success. With the smart card, it has proved difficult to convince the players (train, bus and metro companies, parking operators and banks) of the advantages of the new system. The new system requires new efforts from them (it involves switching costs and investment in new equipment). The ultimate aim is to apply the technology nationwide, which implies that even more actors need to become convinced of the advantages of the new system. Some important actors, such as the SNCF and the national transport ministry, already back the system.

A complicating factor is the new way financial flows are to be managed within the new system. With the old (magnetic) system, the dedicated information system managed the repartition of revenues and costs amongst the organisations involved. For the new system, the plan is to involve banks for the financial aspects, as this is their core competence. Some major French transport companies (SNCF, the Paris metro operator and the Marseilles Department of Transport) are now involved in talks with banks to develop a card that includes an electronic purse. In the long run, the system can be used for other than transport purchases. This example shows how the introduction of ICT may lead to entirely new combinations and possibilities in other fields.

Transport: A Multi-modal Traffic Information Server

ICT may help to improve the information on the traffic situation, which may be of great help for citizens to make better informed choices. Also, it enables a more efficient use of existing infrastructure. In Marseilles, to improve the information provision on the traffic situation and the possibilities of public transport, a partnership of metropolitan actors is now developing a central server that should process this information and make it available to citizens via the Internet and Minitel. The aim is to promote the use of public transport and ensure a more efficient use of the road network in the agglomeration.

Partners in the project are the transport companies in the region SNCF (the national train operator) and RTM (Marseilles public transport), operators of the road network and the City of Marseilles, the adjacent pays d'Aubagne, and the Bouches-du-Rhône *departement*. Each partner provides information to the central server named 'Lepilote', which is managed by an organisation of the same name. Table 9.1 shows the type of information that is delivered by the different partners. Figure 9.1 shows the providers of content, the management of information and the distribution to end users. The server contains a route planner that generates optimal public transport connections from a to b.

Strategic Issues

A notable strength of this service is the integration of road and public transport information. In a European perspective, this is an innovative feature. For citizens, the combination enables a more informed and rational choice of means

Table 9.1   Lepilote: type of players and information

Type of information	Public transport companies	Road networks operators	Public agencies
Permanent	Lines, timetables, tariffs	Advice	Public places
Temporary	Temporary changes	Maintenance	Events
Real time	Delays	Accidents, congestion	Incidents

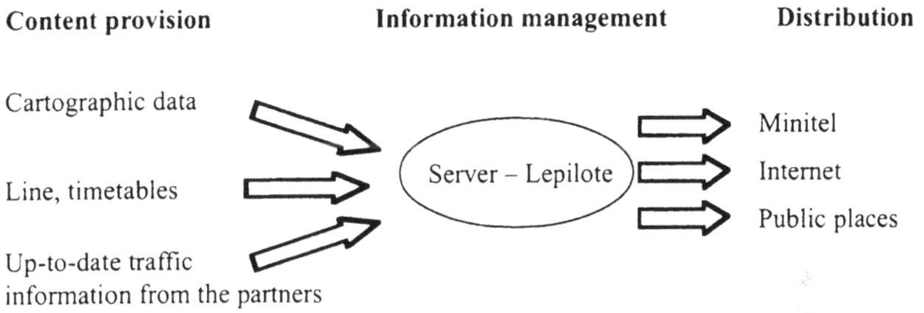

Content provision   **Information management**   **Distribution**

Cartographic data

Line, timetables   Server – Lepilote   Minitel
                                                      Internet
Up-to-date traffic   Public places
information from the partners

Figure 9.1   Information: content, management and distribution

of transportation, resulting in reduced travel time or lower costs. For the city, the system could reduce congestion and thus increase accessibility.

For the system to really add value, it is important that the real-time features are functional, which is currently not the case. This will probably be a question of time. A critical issue is the number of partners involved. Currently, there are only two transport companies, the SNCF and the RTM. The bus lines are not included. To really add value, all the transport companies active in the metropolitan region should be included in the partnership. Another issue is whether the Lepilote server should remain a Marseilles project or extend to other cities in the region of in France. In time, the real-time traffic information could be used to inform drivers on the road with the help of flexible traffic information points.

Municipal Website

Marseilles's municipal website is an example of the city's 'progressiveness' in the information age. Recently, it won the first prize in a French contest for best municipal website regarding content provision and scored second for interactivity. The site attracts 18,000 visitors monthly. It addresses three main groups: visitors, inhabitants, and companies. For visitors, the site contains information on events, cultural provisions, guided tours, local travel possibilities, etc., but also virtual tours. Targeted searches can be carried out with a search engine. Online booking (of hotels for example) is not yet possible; more generally, the interactive features of the tourist site are very limited. For citizens, the site offers additional information on large infrastructure and construction projects that are carried out in the city, on welfare programmes, on childcare possibilities and much more. It also provides help with finding the right municipal department for specific issues and questions. Recently, it has become possible to order official notifications (on marital status and date/place of birth) directly via the Internet, by filling in an online form. More interactive features are to be added (see http://www.mairie-marseille.fr/).

Of course, not all information should become public. The choice was therefore made to create a hierarchy in the access to the information. An intranet was created for civil servants, where much information is (or will soon become) available.

Organisational Issues

The information that is put on the intra- and Internet was already available. Marseilles operated an extensive and detailed geographical information system with all kinds of data. Now, the city has managed to organise the information in a coherent way. The city's Department of Information plays a central role in the transformation to web-based information and service provision. It collects, stores and manages information of all the municipal departments, and thus is the central spider in the city's information web. In the development process of the website, cooperation between the information department and the communication department (responsible for communication between city and citizens and companies) was crucial. A few people from these two departments have made the site tick. For a site to be successful it is crucial that it contains up-to-date information on all municipal affairs. To achieve this and ensure coherence within the site, an 'Internet-correspondent' was appointed in every department, responsible for information provision for that department. Every month, the correspondents take lunch together and discuss site-related issues. The ambition of the webmaster is to provide as many services online as possible, and to ensure optimum access, not only for PC owners but also through interfaces in public places. The last issue is pressing, as at the time of writing only 23 per cent of the Marseilles population has access to the Internet.

The CITIES Initiative: An Integrating Service Platform

In 1994, the cities of Rome, Madrid, Brussels and Marseilles started an experiment to put content on the world wide web in an EC-sponsored project named MIRTO. Later, it was decided to go a step further and offer public services via the web. In this project, the city had Olivetti and Alcatel as industry partners. After expiration of the MIRTO-project, the four cities decided to continue and deepen their efforts in the so-called CITIES project (Cities Telecommunications and IntEgrated Services), within the fourth framework EC programme.

The objective of the project is to establish a multisectorial generic service delivery infrastructure for the Metropolitan Areas of Rome, Marseilles, Madrid and Brussels. The idea is to create a uniform platform that integrates all kinds of services to the population. People will no longer need to memorise different connecting procedures to get information or services on a specific topic: with CITIES they will get the 'one-stop shop' that they have always expected. Whatever the domain is, the answer will be in CITIES – or, at least, this should be citizens' perception.

The variety of services already offered and the large scale of targeted user communities provides a sound basis for validation of the integrated services. Services delivered in CITIES are based on services developed and validated in the framework of previous projects. The project will bring together and reinforce several European projects.[1] From the technology point of view, the CITIES integrated system uses state-of-the-art telematic infrastructures and focuses on the most innovative aspects of each of the contributing projects.

At the time of writing, the platform is ready. It is a general interface to which different servers can be linked functionally, while at the same time staying independent. The interface is to be completed with a intelligent search engine that should turn the web into a very customer-friendly product. This search engine should be able to provide answers to multifaceted questions: for instance, if you look for a cinema show in the city, you would get not only all the movies available at that time, but also information on tickets, online ordering possibilities, suggestions for (public) transport connections to the cinema, including real-time information on delays, etc. The ambition of CITIES is to have a platform that can answer all questions, by 2003. The aim is to develop the maximum number of access points throughout the city.

An important aspect is that CITIES integrates and coordinates all the experiments with ICT within the Marseilles community. This prevents the final user (the citizen) from having to deal with a multiplicity of systems and technologies. CITIES is more than a technical platform: it is also a network to discuss all kinds of obstacles and problems of legislative, administrative and organisational nature that arise in the application of ICTs in public services. The CITIES platform is used to discuss and solve common problems, notably of security and identification: for each of the participating organisations, it is extremely important that the information cannot be hacked into and leaked to the wrong people, and that online payment is secure and the services can only be accessed by those with the right to access (identification). Within the CITIES platform, an interface that is secure and allows payment is being developed.

The city of Marseilles and its partners have invested EURO 1,542,130. An additional EURO 1,071,250 has been invested by the industrial partners. The EC provides 39.67 per cent of total costs (www.cities.irisnet.be).

The CITIES portal currently connects and combines four servers: on transportation (the Lepilote system, see last section), on health (exchange of patient information), on education (web of basic education institutes) and on public services (the Internet-site of the city). The search engine is able to answer complex questions combing information in these fields.

After expiry of the Mirto project, the industrial partner Olivetti wanted to sell the web technology to the participating cities. Marseilles refused, arguing that the city had done most of the work and controlled the content without which the web system would be useless. In the new project, cooperation with Olivetti was abandoned and France Telecom came on board, albeit under strict conditions and with room for the cities to chose another partner eventually.

4   Conclusions

This case study shows how the use of ICT can contribute to the attractiveness of a city in several respects: it can improve the quality of municipal services to citizens; it may help to fight congestion, improve the use of existing infrastructure and thus make the city better accessible. More generally, it helps to increase the quality of the urban product itself, as well its visibility and accessibility.

Often when introducing new information technology, a critical success factor is the number and type of partners who join. Positive network effects are at work, as the chances of success of a system are greater the more actors cooperate. This is clear in all the projects mentioned in the case study. The introduction of the magnetic card, for instance, was so successful because so many actors decided to participate.

A 'network champion' who pushes the new technology and takes the lead in its development and implementation, can do much to speed up adoption by other actors. In this case, the strong position of the municipal transport department (RTM) greatly contributed to the adoption of the magnetic card. In the CITIES project, the creation of a platform to integrate services is clearly led by the four participating cities. It follows that urban organising capacity – the ability to enlist and involve all relevant actors – is a critical capability of the municipal organisation in the information age.

A related issue concerns agreement on standards. The introduction of new information technology often requires such an agreement. To get a standard accepted (which can be a smart card of a certain specification, or the architecture and layout of the municipal Internet site, or a platform for integrated services), its adoption by a few leading players can convince smaller ones to join the bandwagon. If important players do not accept the new system – which can happen when they face high switching costs, or are not convinced of the superiority of the new technology), it is doomed to fail.

Introducing a new system can be a trigger for actors to cooperate and can be a stimulus for even further cooperation. The 18 bus operators on the track Marseilles–Aix-en-Provence now cooperate because they share the new information system. In similar vein, within the municipality, the adoption of the Internet as the central medium for information provision and service delivery can trigger intra-municipal cooperation and improve the general level of service provision even further.

Innovation often thrives on the efforts of a few people. The website case shows that a few enthusiastic and visionary people in a strategic position in the municipal organisation can be decisive in the move of a city towards interactive public service delivery to citizens and e-government. Marseilles would not have been elected best Internet site without these people.

For cities that adopt Internet, it is crucial to debate, define and control the degrees of access to information for different users. Civil service officers need access to other information than do citizens, or companies. Marseilles's solution, with an intranet for internal use, and an Internet site targeted at different user groups (citizens, visitors and companies), is clear and seems to work reasonably well.

The Internet offers the possibility of combining information from various sources in real time, with enormous potential added value for users. Examples are the provision of real-time traffic information on a website. Again, to reap the fruits of the new technology, (municipal) organising capacity is crucial. In this case, the central ability is to convince all the different (public and private) content providers to bundle and regroup their information flows at one point or for one group. The Lepilote project is a courageous effort to bundle information (both permanent and real time) on road traffic, public transport and all kind of tourist information from different sources, to create one product that is extremely useful for the customer/end user.

In the short and medium terms, the role of Internet as a medium for information provision and service delivery should not be overestimated, because Internet penetration among the population is still limited. Also, in many instances, the PC is not the appropriate medium to receive information, for instance when you are on the road. Additional media should be used for public information provision. Marseilles's use of signboards at transport nodes and information pillars is a good example of information distribution by other media.

For cities, the risk is great to rely much on (or even give the lead to) the technology suppliers. To make technology an instrument instead of a goal in itself, urban managers should first and foremost have clear ideas on what they want with the new system, and under what conditions, and then let the supplier come up with appropriate solutions. To avoid being 'locked in' by a single telecom operator or software house, cities should be very careful not to bind themselves to a specific system designed by a single supplier. They should be careful in contracting and ensure escapes to other suppliers in case of exceptional price charges or under-performances. Organising competitive bids including service and maintenance contracts can be a means to elect the best supplier.

The city of Marseilles has benefited greatly from EC support in its ICT projects, particularly within the fourth framework programme. Examples are the Stradivarius project (the smart card), the MIRTO and CITIES projects. Very interestingly, in the CITIES project the participating cities have played a dominant role in these projects, compared to the industrial partners. This is a remarkable break with the past, when the EC (notably the DGs 13 and 12) offered the lead to industrial partners (e.g. software developers or telecom operators). Only managers from these companies were assigned as permanent experts in the EC programme commissions that developed and monitored the programmes.

In sum, as *The Economist* (2000) put it, organisation, not technology, is the main issue:

> The potential [of e-government] is enormous, but governments will need committed leadership, a full understanding of e-business principles and a clear strategy for overcoming the barriers to change: the departmental rivalries, the hostility of unions, the fears of

individuals and the sheer size of the thing. For once, the technology – although crucial to making it all possible – is the least of the worries.

The success of CITIES will depend on the quality of the technology (that is, the power and accurateness of the search engine), the number of actors that participate and put content on the server, the quality of the information they provide, the marketing of the service and the availability of access points to the server. The municipality should work on all dimensions to make the platform a success.

For CITIES to gain credibility, to speed up adoption and use by the population, it is essential to increase the number of services that are integrated in the platform. If it takes too long before citizens really get access to public services, they will turn their backs on the system. This implies that much effort has to be put in involving as many partners as possible and as soon as possible.

Issues of security and payment are critical to the development of telematic services to the population. Banks and insurance companies have great experience in these domains. From this perspective, it is surprising that these types of company currently are not partners in the CITIES project. To speed up the solution of problems in this field, their expertise can perhaps be better used.

Currently, four cities are involved in the CITIES project. However, virtually every European city is dealing with the problem of how to integrate information services and standardise and improve their distribution to the population. Therefore, as soon as the platform has proved its success, its technology should be more widely offered. For CITIES to have real European added value, not only the technical aspects should be diffused to other cities, but also the organisational aspects (how to bring the content providers together, how to convince them to join forces, how to reach as many people as possible by creating access points and so on).

Challenges and Questions

The introduction of a single 'front office' – a website offering information and online services – requires smooth cooperation in the back office (the different government agencies). As long as municipal bodies are involved, the major (or aldermen) can use his/her (or their) power and influence to enforce cooperation. When other, external parties are involved as well (such as transport companies) things get complicated. A critical issue is how to involve them.

Electronic service provision blurs the frontiers between public and private domains. Who is responsible for the information that is offered at the 'one-stop shop'? This is a complicated matter when the services combine and deploy data that are collected, stored and managed by different (public and private) organisations. Ownership of information and liability are treated very differently by the law for public and private actors. If a citizen or company is wrongly informed, which law holds? Who is responsible?

Who should run an integrative platform like CITIES? Should it be the city (because it offers so many of the services offered and produces most of the information)? A disadvantage is the bureaucracy and lack of responsiveness to new technological developments Also, the high costs are carried by the tax payers ultimately. Or should it be run by a private actor, such as a telecom operator (because they have the technological competencies)? This would require very strict control on the part of the municipality, to safeguard quality. Perhaps it would be better to develop a public/private setting to safeguard the quality and integrality of the services on the one hand and enable flexible and fast decision-making on the other. Much depends on the revenues. If the system is a success (many people using it), it could finance itself by advertisement revenues.

Note

1   Examples are MIRTO (AD 1011), REMEDES (HC1043), CONCERT/STRADIVARIUS (TR1013), ISAR-T (HC 1027), ATTACH (UR1001), MAGICA (IE2069), MIRTI (TE2008), CAPITALS (TR1007), and several other European such as COPYSMART (ESPRIT4 n°20517), and MIDAS-net (INFO2000).

References

CITIES (1998), Project Programme version 1.1.
Damlamian, A. (2000), *Marseille, Capitale des Nouvelles Technologies de Communication*, Ville de Marseille.
The Economist (2000), survey on government and the Internet.
Ville de Marseille (2000), *Marseille Economie, Bilan d'étape, Année 1999*, Direction Generale du Developpement Economique.

Web Resources

www.cities.irisnet.be/.
www.Encarta.com.

www.mairie-marseille.fr/.
www.lepilote.com/.

Discussion Partners

Mr P. Argence, Centre de Communication Interactive et de Multimedia pour l'Enseignement.
Mr J.C. Aroumougom, Mission des Programmes Privés et Européens.
Mr G. Coquet, Régie des Transports Marseillais.
Mr A, Damlamian, Direction des Nouvelles Technologies.
Mr B. Giraud-Heraud, Association Formation Professionelle pour Adultes.
Mrs C. Laviolette, Mission des Programmes Privés et Européens.
Mr M. Marchi, Centre d'Etudes Techniques de l'Equipement Méditerrannée.
Mr B. Pinori, Association Régionale d'Assistance Respiratoire à Domicile.
Mr A. Schudel, Mission des Programmes Privés et Européens.
Mr A Tort, Téléservices à la Population.

Chapter Ten

The Hague

1   Introduction

The city of The Hague has the ambition to become a strong centre of ICT and telecom activity. In this case study, we analyse the state of the art of the ICT cluster in the region: what ICT-related actors can be discerned in the Haaglanden region, how do they interact and what are their location considerations? Also, we asses and judge the policies and ambitions of the different public players in the region. We deliberately take the regional level, not just the city of The Hague, as our starting point in view of the density of the area (the municipalities actually form one urban conurbation), the many interrelations that exist among the municipalities and the fact that for an ICT cluster, municipal borders are no restriction.

The case study is based on a literature study and a number of expert interviews with a variety of key people in the ICT sector in the region. It is organised as follows: we start with a general examination of the impact of ICT development and application on urban regions. Next, we discuss the issue of clusters and cluster policies, given The Hague's ambition to develop such a cluster. Section 3 describes the state of the art of the ICT cluster in the city of The Hague and the wider region. Section 4 takes a policy perspective: it describes and judges ICT-related policies in the Haaglanden region. Section 5 draws conclusions and generates ideas to overcome shortcomings.

2   ICT and Urban Development: Some Theoretical Perspectives

ICT has an enormous impact on existing economic activities in urban regions. It makes markets more open and transparent and thus increases competition; it enables firms to become more productive, as they can better organise their production processes and optimise their value chain, for instance with the help of enterprise resource planning systems. Furthermore, a number of new combinations of ICT and virtually all existing branches are emerging. Examples are online-banking services, tele-shopping, digital broadcasting and e-commerce activities. Particularly in the media industry, ICT developments create far-reaching dynamics. One example is the recent merger of America Online (the world's largest Internet access provider) with Time Warner, a 'traditional' media company (among others, *Time Magazine* and CNN). The new combination will be able to develop content and distribute it with both old (TV, newspapers) and new (Internet) media. But in other industries, similar things are happening: examples are online banks (the Security First bank is a success story in the USA); entertainment companies that are starting to develop online products (Disney); and all kinds of new concepts aimed at selling existing products in a new medium (Amazon.com; Dell Computers). In some cases, firms that apply new ICT applications completely reorganise and apply new location strategies that fit their new position. A recent example in the Netherlands is ABN/AMRO, which announced the closure of many of its offices that had become unprofitable because of the rise of online banking.

ICT technology fundamentally changes the role of distance and accessibility, which consequently impacts on urban development. Commuter patterns are changing because of the rise of teleworking. Some activities can be located at different locations compared to the past; physical and non-physical activities of one firm can be located separately. A well-known example is the creation of virtual helpdesks (such as call centres): it does not matter where they are located as long as they can be staffed and are well-connected; another example is the virtual bookstore of Amazon.com, which, unlike normal large bookstores, does not need an inner-city location. Thus, entirely new location patterns emerge; cities need to be aware of this dynamic to pursue successful economic strategies.

ICT as New Growth Sector

ICT developments not only indirectly have an impact by affecting existing branches: the ICT sector itself is an enormous growth sector and thus opens perspectives for new economic activity and employment in urban regions. In Europe, sector growth figures are impressive: see Figure 10.1. The total value of the European market amounts to EURO 365 billion, 5 per cent of the European GDP. The ICT market can be separated into information technology and communications technology; both have a share of 50 per cent, but turnover growth in IT is higher than in communications. The highest growth expectations are in consultancy and outsourcing of IT management (EITO, 1998).

There are several indications that urban regions benefit more than average from the rise of the ICT sector; research suggests that urban regions take the lead in the development and application of ICT software, hardware and

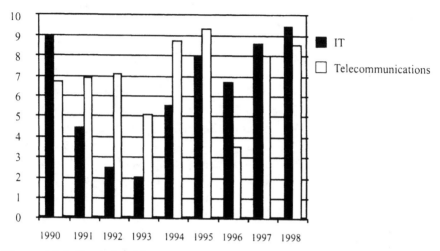

Figure 10.1 Growth of IT and telecommunications market in Europe, 1990–98, annual per cent growth

Source: EITO, 1998.

infrastructure (Abler, 1977; Alles et al. 1994; Graham and Marvin, 1996; Schmandt et al., 1990; Shields et al., 1993). ICT is developed and first applied in urban regions, thanks to the concentration there of many critical users; cities are also the places where new infrastructures (such as high-fibre cables) are first constructed, which further strengthens their position.

Building an ICT Cluster: Notions from Theory

The city of The Hague has the clear ambition to benefit from the growth of the ICT sector and become a strong ICT centre. To find out how this can best be achieved, it is instructive to look at theoretical notions on clusters: how do urban/regional clusters develop and what does that mean for the role of local/regional governments to boost the development of a target cluster?

Krugman (1991) sees only a limited role for the government in cluster stimulation: clusters come into being by chance, not because of policy. He stresses the principle of self-reinforcing cumulative development once a cluster has been established at a certain location: a certain size can be the trigger for take-off, owing to localisation and agglomeration economies: the presence of many similar firms within one industry in a region offers scope for a pool of specialised suppliers, specialised labour and specific education institutes and programmes, enhancing the efficiency of the industry as it generates cheaper and better inputs. In his famous study on industrial clusters, Porter (1999) stresses the role of large and critical local demand as condition for cluster development and trigger of innovations. This demand may be important for ICT clusters as well: there is some evidence that ICT firms tend to locate near their main customers, in regions with high concentrations of financial and business services (Roost, 1998). In a large comparative study on growth clusters in European cities, Berg, Braun and van Winden (1999) take an integral perspective: they find that the growth potential of a cluster is not only determined by *cluster-specific conditions* (such as the quantity and quality of the firms, strategic interactions amongst them and the creation of new firms), but also by *the general spatial-economic situation of the urban region* (economic structure, quality of the living environment and accessibility). Finally, they state that the degree of '*organising capacity*' matters: regional leadership, an integral strategy on the cluster, constructive public/private cooperation and sufficient societal and political support are additional necessary conditions for the successful development of a cluster.

Policy Options

This last approach gives direction to policy-options that are open to local and regional governments to boost their ICT sector: strategies can be designed to: 1) improve general conditions; 2) increase the organising capacity of the region; and 3) improve ICT cluster-specific conditions.

Improve general economic and spatial conditions For an ICT cluster, highly skilled people are badly needed. To retain or attract these people, investing in the attractiveness of a city in terms of high-quality housing, cultural and leisure facilities is important. Also, general policies aimed at strengthening the regional economy may indirectly be beneficial for the ICT sector, as a stronger economy extends the sector's client base.

Strengthen organising capacity This can be achieved by developing a regional vision and designing an integrative strategy for the development of the cluster. This requires leadership, given the multitude of interests. Organising capacity can further be strengthened by involving cluster actors in policy making processes, to avoid policy mismatches and unrealistic strategies.

Improve cluster-specific conditions, aimed to strengthen the cluster directly. First, local governments can create good conditions by offering appropriate office space and contribute to the electronic infrastructure in the region. To that end, it is important to have a good insight into the location preferences of several type of ICT firms. Second, it can help to strengthen strategic networks among ICT firms and between ICT firms and the education infrastructure, to make better use of the local resources. Third, it can play a role in supporting new firms – a very important source of dynamics – in the field of ICT, by providing venture capital, cheap office space, or linking them up with potential clients. Fourth, local government can stimulate the ICT sector indirectly, by acting as a critical demander of ICT services, or by helping SMEs in the region to apply and adopt ICTs, for instance by offering computer courses, or subsidising consultancy days for the implementation of ICT in SMEs. This enlarges the local market for the ICT sector. A final option is to actively attract ICT firms from outside the region. Here, direct competition with other cities that pursue similar strategies may emerge. To prevent unproductive and costly competition between neighbouring cities, careful policy coordination on the appropriate regional scale is necessary.

3   An Image of the Region's ICT cluster

After this theoretical introduction, this section becomes more concrete: it describes the components and dynamics of the ICT sector under discussion. First, it introduces the city of The Hague and the region of Haaglanden. Next, it elaborates on the private and (semi-)public players in the ICT sector and finally, it looks at strategic cooperation in the cluster.

The City of The Hague and its Region

With 445,000 inhabitants, The Hague is the third largest city of the Netherlands and is the seat of the Dutch national government. It hosts virtually all the national ministries, as well as foreign embassies and other diplomatic functions. The city is well known internationally as the seat of the International Court of Justice. Functionally, the city forms part of the Haaglanden Region, which also comprises the cities of Delft, Zoetermeer, Voorburg, Rijswijk, Leidschendam and some smaller municipalities. Together, these municipalities form one large urban sprawl.

The economy of the city of The Hague is predominantly service oriented: administrative services but also all kinds of (international) business services such a publishers and insurance companies. Tourism is important as well; the city attracts many visitors and tourists to its beaches, its city landscape and its cultural attractions such as the Mauritshuis. The economic structure of the city is depicted in Table 3.1

Table 10.1   Employment in several clusters

Clusters	% of employment
Administration and international organisations	22
Commercial services	20
Communication	7
Shops and leisure	17
Knowledge and education	6
Health services	15
Small manufacturing and wholesale	10
Rest	4
Total	100

Source: NEI, 1998.

The general economic strategy of the city of The Hague is further to strengthen the image of the city as an international city of law and justice, and to stimulate entrepreneurial activities in the private sector; the city not only wants to be known for its administrative role, but also as a buoyant centre of business. The city invests heavily in high-class business locations (the Hoog Hage project), improvement and restructuring of the housing stock to attract highly-skilled inhabitants and stimulates tourism development. Most urban investment (more than EURO 2.3 billion) is concentrated in the centre of the city, where the skyline has been transformed in the last five years.

The ICT-cluster in the Haaglanden Region

The city's ambition to become a strong ICT/telecom city forms part of the strategy to develop more (international) private business. The Hague region can already boast a strong ICT sector. The Hague counts several headquarters of telephone service providers, among which are the headquarters of Dutch market leader KPN (fixed and mobile phone services, infrastructure). This firm has made the Hague its home for many years. Recent newcomers in the city are the mobile communications service providers Ben and Dutchtone. The latter are new entrants (since 1998) in the liberalised Dutch telecom market. They are owned by foreign companies: Ben is a combination of

Belgacom and Tele Denmark; in 1998, it had 320 employees in The Hague, but this number is rapidly increasing. Dutchtone is a subsidiary of France Telecom, and has a staff of around 550 in The Hague (City of The Hague, 1999).

Both are aggressively seeking to gain market share and expanding very rapidly in terms of turnover and employees. The companies located in The Hague because of the pool of qualified labour (thanks to KPN and the good public knowledge infrastructure), and the international image of the city, with its high concentration of diplomatic activity and international schools. It should also be noted that acquisition activities of the city of The Hague have contributed to attract these firms (more on this issue in the next section). Some claim that the presence of the Dutch telecom regulation authority (OPTA) in the Hague has also played a role in attracting these firms.

Ben, a new telecom operator in the City of The Hague
Until the early 1990s, there was only one telephone service provider in the Netherlands: the formerly state-owned KPN. Since the liberalisation of the Dutch telecom market, four new players have entered, particularly in the rapidly expanding market of mobile telecommunications. One of them is Ben, a company owned by Belgacom, Tele Danmark and Ameritech. Ben seeks a unique selling point by offering simplicity, clarity and a very personal approach in the market of mobile services. In 1998, the firm located its headquarters in the Laakhaven area, a business district in the city centre. The Hague was considered a good location, among other reasons because of the knowledge infrastructure and the presence of specialised staff.

In the slipstream of these telecom providers, The Hague hosts technical equipment firms such as Alcatel, Siemens and Nokia, mainly functioning as sales outlets and customer services centres, not R&D facilities. Other well-known players are Amazon.com, the American e-bookstore, which has recently opened its European headquarters in the centre of The Hague (staff: around 400), serving as the European call centre and logistics control centre for the European market. Another North American Internet firm in the Hague is Mapquest, a leader in interactive geographic information systems. These companies wanted to locate in The Netherlands anyway, but choose the city of The Hague for its international atmosphere and the availability of (ICT) staff.

Amazon.com: European headquarters in The Hague
The world's largest online retailer, Amazon.com (based in Seattle), recently opened its headquarters/customers service centre in The Hague, offering 300 to 500 jobs. This firm offers online 4.7 million products, mainly books, CDs, and videos, has 12 million clients worldwide, and generates a turnover of EURO 910 million (*Haagse Courant*, 14 October 1999). Ordered products are delivered at home within 24 hours (in the US). The location of Amazon.com in The Hague fits in its strategy to become stronger in the European market: a presence on the European continent is felt necessary to keep in touch with consumer tastes, and realise faster delivery. From its location in The Hague, the firm processes all European orders and invoices, answers customers' questions, and organises the distribution. The actual physical distribution of the ordered products for Europe takes place from centres in England and Germany. Amazon.com considered the Netherlands in general an attractive base, for reasons of tax regulations, culture, and accessibility. The Hague proved attractive because of the presence of sufficient multilingual staff and its international image.

Apart from these well-known big international companies, the cluster consists of a rich diversity of smaller ones: software developers, multimedia firms, CD-ROM producers, IT consultants and so on. They benefit from the presence of many clients in and around The Hague. For many of these firms, the urban atmosphere is an important consideration in locating in The Hague. Some of these firms are concentrated in The Hobbit, a privately-owned business centre for IT and telematics: see box.

The neighbouring cities of Delft and Zoetermeer also host a lot of ICT activity. A certain specialisation of ICT activities can be observed: in The Hague, headquarter functions of telecom firms dominate the scene, although (new) media firms and small ICT firms are also present thanks to the 'urban atmosphere' and the presence of so many clients. Zoetermeer is particularly strong in software and IT consultancy companies: its relatively good accessibility makes it a good location for these firms, which frequently interact with clients all over the country. Delft, home of the large Technical University, has a 'natural' attraction as an R&D centre in the field of ICT: it hosts a rich variety of small and medium-sized ICT firms, some of them concentrated in commercial business centres such as RADEX and the Business Technology Centre.[1]

> *The Hobbit: building synergy in ICT and telematics*
> The Hobbit is a multi-company building for ICT-related business. It is
> located in an old switchboard installation building (with the advantage
> of good external data connections) and was redeveloped in 1995 by a
> private project developer. It hosts 33 small firms (total employment:
> 200), among which are not only software developers, Internet designers,
> Internet access providers, but also nontechnical ICT-related firms, such
> as lawyers specialising in intellectual property issues, a publisher of an
> IT magazine and IT consultants. From the beginning this developer
> aimed at realising synergies and cooperation in the building. Therefore,
> tenants are screened for their value-added for the concept. A facilities
> provider in the building, Hobbit Facilities, plays a central 'bonding'
> role in the concept: to stimulate cooperation, it organises theme sessions,
> sports events and, for instance, an organised tour to CEBIT, the major
> computer fair in Europe. The concept seems to work well: there is a
> waiting list for tenants and the number of bankruptcies is very low
> compared to the national average. The role of the city of The Hague
> has been limited: it only offered rent subsidy for firms during the first
> year the building was in operation.

Knowledge and Education Infrastructure

A very important asset of the Haaglanden region (according to the firms we
interviewed) is its strong education base: see Table 10.2. The Haagse
Hogeschool, an institute of higher education with 15,000 students, of whom
2,600 specialised in informatics, is the largest supplier of IT graduates in the
Netherlands. The yearly outflow of graduates provides a crucial influx of
qualified staff for existing ICT firms in the region. The school also constitutes
a breeding ground for new firms. The nearby Technical University of Delft is
well known internationally; this institute also 'produces' a substantial outflow
of highly skilled staff, and generates several spin-offs.

Is it a Real Cluster? Cooperation among the ICT Sectors

The last sections have shown that the region counts a number of ICT firms
and knowledge institutes, but to what extent do they together constitute a
real, fully-fledged cluster? What is ties the cluster together? One conclusion
is that the large pool of specialised labour (particularly in the field of telecoms)
is an important engine behind the development of the cluster. The excellent

Table 10.2 Higher education in ICT

	Total number of students	ICT students	Annual ICT graduates
Haagse Hogeschool	15,000	2,600	700
Delft Technical University*	13,000	1,850	700

*   Faculty of Information Technology and Systems, 1998.

Source: http://www.its.tudelft.nl/.

educational and research infrastructures in the region are key assets in this respect. The attractiveness of the region as a living location also plays a role: it makes it easier to attract staff from outside The Hague.

We found that strategic inter-firm cooperation is not characteristic of the ICT business in The Hague: most firms work on a 'stand alone' basis or are mainly oriented to other establishments of the same firm. In The Hobbit there are some interesting joint projects. The large telecom firms in the region – competitors – frequently meet to arrange matters of common interest such as telephone number portability. They also have strategic relationships with equipment providers in the region (Ben is oriented to Nokia, Dutchtone to Alcatel). Siemens is active with several partners in many projects throughout the Netherlands. Syntens helps firms to get into contact with partners and is particularly active in coupling existing firms to ICT firms, to establish new combinations.

The main education institutes are increasingly open to interaction with their environment. The Informatics Department of the Haagse Hogeschool interacts with local and regional businesses; not only do students take on apprenticeships, the institute also organises courses for staff of firms – KPN staff give guest lectures; there is a link with KPN research. The Haagse Hogeschool is also involved in a think-tank of the city of The Hague on the information society, the Resident.net project and cooperates with the acquisition organisation to show the education infrastructure in the region to interested firms. The Technical University of Delft (Faculty of Information Technology and Systems) executes more and more contract research for ICT firms, albeit not specifically at the regional level.

4   ICT Cluster Policies in the Region

The city of The Hague has the clear ambition to be a ICT/telecom city in the Netherlands, as can be read in several policy documents. What policies are being pursued to meet this target? In this section, we not only take the actions of The Hague into account, but also look at cluster-specific policies in the entire Haaglanden region, by several type of actors: public (municipalities); semi-public (intermediary organisations and knowledge institutes); and private actors. Four types of policy target are depicted in Figure 10.2: acquisition of (foreign) firms; starters' support; infrastructure policies; and ICT-adoption stimulation measures.

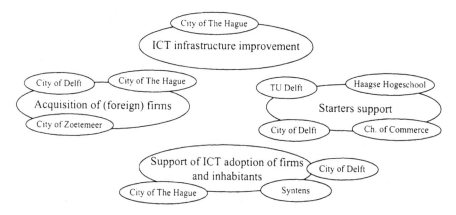

Figure 10.2 ICT policies in the region

Acquisition of Foreign Firms

In the region, The Hague and Delft seek to attract foreign ICT firms. The city of The Hague focuses on the USA and Canada, with a special focus on call centres and mobile communications firms (based on a market study by M&I that identified these branches as fast-growing and interesting targets for The Hague). For that purpose, it cooperates with the The Hague Development Foundation, a private organisation that tries to increase business activity in the city. Potential 'clients' are actively spotted, and eventually helped to find an appropriate location, staff and advised on all kinds of legal aspects. In the period 1998–March 1999, six firms were attracted, offering a total employment of 1,500 (Interim Report on Acquisition Results). Some of the most notable successes are the location of Amazon.com and Mapquest in The Hague. Some acquisitions failed, mainly because of the lack of 'ICT image' of The Hague, strong competition from Amsterdam and Rotterdam, the absence of a university, and the lack of market communication.

The city of The Hague competes not only with the other large Dutch cities (Rotterdam, Amsterdam, Utrecht) but also with municipalities in the region, which in some cases have their own, separate acquisition activities. The city of Delft for instance seeks to acquire high-tech firms from Israel. The city hosts (and co-finances) the Israel Office Delft, offering assistance in renting small units in high-tech buildings, registering patents, networking with high-tech companies, consulting organisations for subsidies and finding administrative staff free of charge.

Up to the time of writing, there has been no satisfactory coordination in acquisition, resulting in a waste of resources and creating a source of conflict. This problem is recognised by the cities concerned. Therefore, a new acquisition organisation is being set up, in which the cities of The Hague, Zoetermeer, Delft and Leiden (not part of Haaglanden, but located nearby) and the regional Chamber of Commerce should cooperate in joint international acquisition of business. The smaller municipalities are not taking part. The idea is to let each city focus on a specific branch, i.e. The Hague on telecom companies, Delft on high-tech developers and research institutes, and so on, but individual cities should still be allowed to give individual financial incentives. Many difficulties have been encountered in setting up this new organisation: a severe lack of mutual trust – at political and administrative levels – among the partner cities is the principal cause.

Starters Support

Entrepreneurship is stimulated by all the municipalities and many other (semi-)public actors in Haaglanden. Many initiatives have been taken to facilitate new business start-ups, some specifically aimed at ICT starters, others more general:

- the Technical University of Delft encourages graduates to start up new businesses; ex-students are allowed to use university facilities (labs, computer networks) and are offered 'soft' loans. Many start-ups are located in an obsolete building of the Technical University (it is unclear where these firms will move to when this building is demolished);
- the Haagse Hogeschool operates a centre (CISOH: Centrum voor Innovatieve Ondernemingen Haaglanden) to enable entrepreneurial

students to run their business. For one year after their graduation, they can use facilities and spaces in the Haagse Hogeschool. Ten successful firms left the centre in 1999;

- the city of The Hague does not consider support for starters as a primary policy goal: it is left to banks, investment funds and Syntens, although The Hague participates in a foundation named 'StaBij' to support starters. It has also provided rent subsidies to firms in The Hobbit;
- the city of Delft (Delft Kennisstad) actively engages in ICT starters support. For a few years now it has operated a foundation for starters, but this initiative has not been successful. Currently, it has concrete plans to set up a twinning initiative together with the city of Rotterdam. In The Netherlands, the twinning initiative was launched by the Ministry of Economic Affairs, with the aim to raise more world class ICT firms in the Netherlands.[2] Twinning centres are already present in Amsterdam, Eindhoven and Twente. The city of Delft also submitted a proposal to participate, but it was rejected. Currently, Delft is preparing a second try together with Rotterdam, with a 'hub and spoke' concept, dubbed IC/ICT (International Centre for ICT). This deviates from the existing twinning centres in two basic respects: first, the requirements for firms should become less severe, and second, firms need not be located in one particular building. This offers more flexibility for participating firms. The management and support (the 'hub') should be organised from one central location. Although the initiative has been taken by Delft and Rotterdam, firms (or business buildings such as The Hobbit) from The Hague are approached to connect to the concept.

Stimulation of ICT Adoption

Local governments throughout Europe are trying to prepare and guide their citizens and firms – mostly SMEs – in the information society. Likewise in the Haaglanden region. Several actions have been taken and projects initiated to stimulate the use of ICT:

- *Helping SMEs to use ICT.* Syntens – initiated and sponsored by the Ministry of Economic Affairs – is particularly active in this field. It is a national organisation, with 15 local establishments, whose general objective is to help firms to innovate. In the establishment of Haaglanden (responsible for the northern part of South Holland, including Delft), 24 consultants are active in connecting firms to each other and to knowledge institutes, and consult firms in their innovation processes. Syntens has the impression that SMEs have much difficulty with the transition to an information economy. A concrete project in the field of ICT is therefore aimed at stimulating e-commerce and website development of SMEs. Firms get a few days of free consultancy. This also opens new business opportunities for ICT firms, who see their client base increase as a consequence.

- *Stimulate citizens to use the Internet.* A vast majority of the population is still not online. To increase the use of the new media and prepare its citizens for the information society, the city of The Hague has started the Resident.net, in cooperation with Casema (cable operator) and KPN (a telecom provider). Every inhabitant of the city should get free access to a limited part of the Internet. Also, information can be retrieved via interactive teletext; with a telephone and a TV, the Internet can be browsed without a PC. The project also aims to make the many different websites of the city more accessible, by constructing a search engine and creating improved directory structures. The city of The Hague has invested EURO 1.13 million in the project. A critical note could be that the city, although it is the principal funder and initiator, has lost too much control over its project to KPN, the former Dutch National telephone company that runs the operation. Resident.net has been developed as a portal of KPN. KPN will probably gain most from the project: not only will it make money from the helpdesk (which is relatively expensive) and the telephone costs of Internet use (KPN is the main telephone supplier in the region), users of Resident.net also need to supply a lot of personal information to KPN.
- *Improve municipal services.* All municipalities in Haaglanden share this ambition. The city of The Hague wants to use ICT to improve its service level: make it more user-friendly and provide access to all kinds of information that firms or citizens might need. Resident.net is an important project in this respect; the same holds for the Glazen Stadhuis (Glass Townhall) project, aimed at 'informatising' the municipal organisation. This has large implications for the municipal organisation: every department has to find out what its information product is and new combinations of information have to be composed from different departments. The city is also experimenting with interactive policy and decision-making, in cooperation with the University of Leiden.

Infrastructure Policy

High-capacity electronic infrastructure is a necessary condition for much ICT activity. In this respect, the city of The Hague has no competitive advantage or disadvantage over other Dutch cities. In general, the electronic infrastructure is considered very good by American companies that (aim to) locate in Europe (Ash, 1999). The Hague has a city ring, constructed by KPN some time ago. Further, the city plans to inform new providers of telecom infrastructure about the possibilities for investment in The Hague. New network providers could be cable companies (UPC, Casema, MCIworldcom), or telephone operators (Dutchtone, Telfort).

Recently, the municipality managed to convince MCI-Worldcom – a world leader in high capacity data connections – to connect The Hague to their new high capacity line running from Amsterdam to Rotterdam, and introduced MCI to the large potential customers in The Hague.

Conclusion

In conclusion, in the small and densely populated area of Haaglanden, the 'policy density' regarding ICT is very high: many actors – at local, regional and national levels, public and private – are involved, and pursue sometimes overlapping policies. An important omission is the lack of a metropolitan spatial-economic vision on the Haaglanden region. This results in competitive relations among adjacent communities, in acquisition of foreign new firms (not only in ICT but in all sectors) and offering of business locations. Municipalities in Haaglanden are very reluctant to exchange strategic information on firms (re)location, for fear of being 'cheated' by the neighbours who might offer an even more attractive incentive. This is all the more amazing because the different types of ICT firms need different environments (see section 3). Thus, better cooperation might help firms to find an optimum location. A severe lack of trust on all levels among the municipalities is a major barrier to cooperation. The recent seting-up of a twinning centre by Delft and Rotterdam without the involvement of the city of The Hague (but possibly with companies located in The Hague) is a case in point.

5   Conclusions

The Hague and its region have all the necessary ingredients for a top-level ICT cluster. The education and research infrastructure is particularly strong, but the private sector is also well represented, with some key players. This forms a sound basis for further growth of the sector. The increasing attractiveness of the city (thanks to heavy investments in high-class housing,

Table 10.3 ICT policy measures by several (semi-)public bodies

Actor	Type of policy	Instrument	Target region
City of The Hague	Acquisition and support of foreign ICT-firms	Advice on locations, help getting staff	City of The Hague
	Provision of infrastructure	Convince suppliers to construct networks in The Hague	City of The Hague
City of Delft	Acquisition of ICT-firms from Israel	Offer all kind of services to Israeli firms	City of Delft
	Facilitate ICT-start-ups	Seed capital and space provision for promising start-ups in ICT	Cities of Delft and Rotterdam
	Knowledge broker	'Knowledge telephone' initiative, to connect demand and supply of knowledge	
City of Zoetermeer	ICT education	Investment in call-centre academy	City of Zoetermeer
Syntens	Knowledge broker	Coupling of ICT firms to other firms	North–south Holland
Chamber of Commerce	Starters support	Offer starters package	North–south Holland
Haagse Hogeschool	Starters support	Offer space and facilities for starters	Randstad
TU Delft	Starters support	Offer space, loans and facilities for starters	Netherlands

office space and in culture) adds to the potential to attract and retain appropriately skilled people and create the proper urban atmosphere for creative industries like the ICT sector.

The ambition of The Hague is to become an even stronger centre of ICT. However, the translation of ambitions into policy and actions shows some weaknesses. First, the focus is too much on attracting firms from outside; second, the sector's development should be seen in a wider regional context. These points need some elaboration.

Despite recent successes, The Hague's focus on acquisition of foreign ICT firms (call centres, telecom firms) has its restrictions. For one thing, given the current scarcity of specialised staff, the question can be raised whether it is useful to attract high technology firms at all; investments had perhaps better be directed to attracting skilled people instead of firms, by creating an attractive living environment, a lively cultural climate and appropriate housing. The firms will follow.

Apart from this discussion, it would be strategic and effective to complement acquisition efforts with a set of measures to exploit opportunities within the region better, in other words: make more use of use local resources. Several options are open. First, the city could promote strategic cooperation in the cluster, for instance by putting a financial premium on innovative cooperations among ICT firms that engage in joint innovative projects. A second option is to make better use of the 'intellectual capital' of the Haagse Hogeschool, for instance by contributing to its connection to a high-capacity data network. Third, the city could stimulate ICT adoption in specific urban sectors in the city. As an example, much can be gained with ICT application in the tourist sector. The city could organise specific information sessions for small-scale tourist establishments, preferably in an entertaining setting for greater effect, offer ICT courses for small tourist firms, or try to develop a website with all the region's tourist information. The role of the national ministries and other public institutes in the city as producers of information could also be exploited better.

A more integrative approach requires better coordination within the city administration. Responsibilities for the ICT cluster are scattered among several groups within the DSO unit; ICT infrastructure policies are coordinated by yet another department, as are as education and starters policies. The establishment of a municipal 'ICT taskforce', consisting of representatives of all these sectors, could contribute to the coherence and effectiveness of the cluster policy.

An example of proactive urban technology policy is the Resident.net project, offering free Internet access to all inhabitants of the city. This helps to make people from all layers of population familiar with the new possibilities, even if they have no PC at home. An important lesson to be learned is that the municipality should remain the leader in such projects and not give too much away to private partners.

On the regional level, it can be concluded that there is a lack of organising capacity, which is reflected in unfruitful inter-municipal competition to attract foreign ICT firms, among other things; also, there is a severe lack of policy coordination in related fields, such as the supply and development of business locations. A notorious lack of mutual trust among the municipalities makes it hard to find a good solution. This pleads for an alternative regional governance structure that would strengthen the region's organising capacity.

The establishment of a regional development company could be an effective tool, governed by an independent board, which should see to the integral economic development of the region as a whole. Such a company could take responsibility for a better distribution of ICT activities, coordinated investment in infrastructure and structured starters' support.

Notes

1   An ICT building, 3,000 m^2, which hosts national and international companies that operate in high-tech industry, others in obsolete buildings of the Technical University.

2   The concept is to construct excellent buildings with the best IT infrastructure, for very promising ICT start-up firms (which means having potential to compete in world markets), and offer them financial and managerial support, and connect ('twin') them to investors, large potential clients and consultants. Advisers are renowned entrepreneurs who support starters with their knowledge, contacts, and experience. They help in answering questions, writing business plans, advise and support. Twinning has also set up two funds: Twinning Seed Fund (venture capital to ICT start-ups, up to EURO 150,000 over the start-up period) and Twinning Growth Fund (financiers who want to invest in a company can ask the Twinning Growth Fund to share the investment on a 50/50 basis up to a maximum of EURO 900,000 each). Total investment of the Ministry of Economic Affairs amounts to EURO 36.3 million.

References

Abler, R. (1977), 'The Telephone and the Evolution of the American Metropolitan System', in De Sola Pool, I. (ed.), *The Social Impact of Telephone*, MIT Press, London, pp. 318–41.

Alles, P., Esparza, A. and Lucas, S. (1994), 'Telecommunications and the Large City–Small City Divide: Evidence from Indiana cities', *Professional Geographer*, No. 46, pp. 307–16.

Ash, N. (1999), 'Dutch Bank on Brains', *Corporate Location*, January/February, Euromoney Publications.

Berg, L. van den, Braun, E. and Winden, W. van (1999), *Growth Clusters in European Metropolitan Cities: A new policy perspective*, Euricur, Erasmus University, Rotterdam.

City of The Hague (1999), *Den Haag: ICT-groeikern van Nederland.*

EITO (1998), *EITO '99, European Information Technology Observatory.*

Graham, S. and Marvin, S. (1996), *Telecommunications and the City: Electronic spaces, urban places*, Routledge, New York.

Haagse Courant (1999), 'Internetreus naar Den Haag', 14 October.

Krugman, P. (1991), 'Increasing Returns and Economic Geography', *Journal of Political Economy*, Vol. 99, pp. 483–99.

NEI (1998), *Meer vraag naar Den Haag, Analyse van en visie op de stedelijke economie van Den Haag*, Nederlands Economisch Instituut, Rotterdam.

Porter, M. (1998), 'Clusters and New Economics', *Harvard Business Review* No. 6, November/ December.

Roost, F. (1998), 'Recreating the City as Entertainment Center: The media industry's role in transforming Potzdamer Platz and Times Square', *Journal of Urban Technology*, Vol. 5.

Schmand, J., Williams, F., Wilson, R. and Strover, S. (eds) (1990), *The New urban Infrastructure: Cities and telecommunications*, Praeger Publishers, New York.

Shields, P., Dervin, B., Richter, C. and Soller, R. (1993), 'Who Needs POTS-plus Services? A Comparison of Residential User Needs along the Rural–Urban Continuum', *Telecommunications Policy*, Vol. 17, pp. 563–87.

Discussion Partners

Mr F.J. van Bork, The Hague City Administration, City of The Hague, Consultant.

Mr T.L. Brill, KPN Royal Dutch Telecom, Manager.

Mr L.G.J. Draijsma, Department of Urban Development, City of The Hague, Project Manager.

Mr J.J. Duivenvoorden, Department of Urban Development, City of The Hague, Manager, Information Provision and Automatisation.

Mr C. J. van Laren, City of Delft, Project Manager.

Mr C.A. van Loon, Haagse Hogeschool, Faculty of Informatics, Managing Director.

Mr A.H. Rijerkerk, Chamber of Commerce Haaglanden, Director, Policy Advice.

Mr U. Roberti, The Hobbit, Alsumex, Project Developer.

Mr W.E.C. Rutgers, Department of Urban Development, City of The Hague, Manager, Acquisition and Account Management.

Mr H. Versloot, Syntens, Consultant.